T0276214

CAMBRIDGE LIBRARY COLLECTION

Books of enduring scholarly value

British and Irish History, Nineteenth Century

This series comprises contemporary or near-contemporary accounts of the political, economic and social history of the British Isles during the nineteenth century. It includes material on international diplomacy and trade, labour relations and the women's movement, developments in education and social welfare, religious emancipation, the justice system, and special events including the Great Exhibition of 1851.

A Journey through England and Scotland to the Hebrides in 1784

The French geologist Barthélemy Faujas de Saint-Fond (1741–1819) abandoned the legal profession to pursue studies in natural history, working at the museum of natural history in Paris and as royal commissioner of mines. His enthusiam for geology took him in 1784 to Britain, to investigate the basalt formations on the Hebridean island of Staffa described by Sir Joseph Banks in Pennant's *Tour in Scotland* (also reissued in this series). His subsequent account was published in France in 1797, and first translated into English in an abridged form in 1814. This two-volume annotated translation by the well-known geologist Sir Archibald Geikie (1835–1924), prefaced by a short biography of Faujas, was published in 1907. The work is interesting for its social as well as its geological observations. Volume 2 describes the geology and natural history of the Hebrides. On his return journey, Faujas also visits the geological marvels of Derbyshire.

Cambridge University Press has long been a pioneer in the reissuing of out-of-print titles from its own backlist, producing digital reprints of books that are still sought after by scholars and students but could not be reprinted economically using traditional technology. The Cambridge Library Collection extends this activity to a wider range of books which are still of importance to researchers and professionals, either for the source material they contain, or as landmarks in the history of their academic discipline.

Drawing from the world-renowned collections in the Cambridge University Library and other partner libraries, and guided by the advice of experts in each subject area, Cambridge University Press is using state-of-the-art scanning machines in its own Printing House to capture the content of each book selected for inclusion. The files are processed to give a consistently clear, crisp image, and the books finished to the high quality standard for which the Press is recognised around the world. The latest print-on-demand technology ensures that the books will remain available indefinitely, and that orders for single or multiple copies can quickly be supplied.

The Cambridge Library Collection brings back to life books of enduring scholarly value (including out-of-copyright works originally issued by other publishers) across a wide range of disciplines in the humanities and social sciences and in science and technology.

A Journey through England and Scotland to the Hebrides in 1784

A Revised Edition of the English Translation

VOLUME 2

BARTHÉLEMY FAUJAS DE SAINT-FOND
EDITED AND TRANSLATED BY
ARCHIBALD GEIKIE

CAMBRIDGE
UNIVERSITY PRESS

CAMBRIDGE
UNIVERSITY PRESS

University Printing House, Cambridge, CB2 8BS, United Kingdom

Cambridge University Press is part of the University of Cambridge.
It furthers the University's mission by disseminating knowledge in the pursuit of
education, learning and research at the highest international levels of excellence.

www.cambridge.org
Information on this title: www.cambridge.org/9781108071574

© in this compilation Cambridge University Press 2014

This edition first published 1907
This digitally printed version 2014

ISBN 978-1-108-07157-4 Paperback

This book reproduces the text of the original edition. The content and language reflect
the beliefs, practices and terminology of their time, and have not been updated.

Cambridge University Press wishes to make clear that the book, unless originally published
by Cambridge, is not being republished by, in association or collaboration with,
or with the endorsement or approval of, the original publisher or its successors in title.

A JOURNEY THROUGH ENGLAND AND SCOTLAND TO THE HEBRIDES IN 1784

IN TWO VOLUMES

VOLUME II

Impression, Four hundred and fifty copies.

View of the Cave of Fingal, in the Isle of Staffa, one of the Hebrides.

A JOURNEY THROUGH ENGLAND AND SCOTLAND TO THE HEBRIDES IN 1784

BY

B. FAUJAS DE SAINT FOND

A REVISED EDITION OF THE ENGLISH TRANSLATION
EDITED, WITH NOTES
AND A MEMOIR OF THE AUTHOR

BY

SIR ARCHIBALD GEIKIE, D.C.L., Sec.R.S.

CORRESPONDENT OF THE INSTITUTE OF FRANCE

VOLUME TWO

GLASGOW: HUGH HOPKINS

1907

CONTENTS

v

CONTENTS

CHAPTER XXI

PLATES

ix

A JOURNEY

ENGLAND AND SCOTLAND

TO THE

HEBRIDES

CHAPTER I

Departure from Oban for the Island of Mull.—Passage of the Sound of Mull.— Small isle of Fiart.—Druidical Monuments.—Arrival at Aros

MY solitary excursions in the neighbourhood of Oban were scarcely completed, and I had done with putting my observations in order, when the people of the inn announced the arrival of a traveller, who, astonished to learn that there was a Frenchman alone in so desert a place, asked to see me.

He was a young English officer, who had come to wait at Oban for a favourable opportunity of going to the isle of Skye, of which he was a native.

He was called Macdonald and had prose-
cuted his first studies at the Scotch college in
Paris; he spoke the French language tolerably, and was not deficient in information.
His arrival was a happy and agreeable
rencounter for me. I told him of the object
of my journey, and of my approaching departure for the isle of Mull, where some
fellow-travellers were waiting for me to visit
the celebrated Fingal's Cave in the isle of
Staffa.

Mr Macdonald replied, that though his
native country was at no great distance from
that island, and though he had often heard
mention of the cave of the father of Ossian,
his education in France and his travels had
not yet given him an opportunity of visiting
a place so remarkable: but, that if I would
permit him to accompany me, he would
eagerly embrace the opportunity of seeing it
with me; and that he would also have the
pleasure of being useful to me in the country,
as he understood the Erse or Celtic language,
the only one in use among the Hebrides.*

* [It will be observed that the author everywhere speaks of
the group of islands from Skye southwards as "the Hebrides."
This term is now generally restricted to the long chain of

I accepted the obliging offers of Mr Macdonald with all the greater pleasure and thankfulness, as they came from a man of a sociable disposition, and were delivered in a tone of frankness and affability which prepossessed me in his favour: he might besides procure me some facilities in the isle of Mull, where he had several acquaintances, particularly Mr Maclean, for whom the Duke of Argyll had given me letters of recommendation. So we now waited only for the arrival of the vessel, which at length entered the harbour on the night of the 23rd of September.

The crew consisted of no more than two fishermen from the isle of Skye, who were clothed in the fashion of the Hebrides, that is, in the costume of the Scottish Highlanders. Our vessel had neither decks nor rigging; she was of the worst construction; and dragged at her stern a small skiff capable of holding at most only four persons.

The fare was agreed upon, and it was settled that we should set out on the follow-

islands from the Lewis south to Barra Head. If it is used as in the text, a distinction should be made between the western chain or " Outer Hebrides, " and the eastern chain or " Inner Hebrides."]

ing day; but I know not from what caprice,
our boatmen changed their minds, and
wished to remain for three days longer.
It was only by remonstrances and coaxing,
and by giving them a present of two bottles
of rum, that Mr Macdonald could at last
prevail on them to make ready to start next
morning.

We left the shore of Oban, at seven in the
morning of the twenty-fourth.   The sea,
though not heavy, was agitated; the winds
were variable, and the currents at the entrance of the Sound of Mull running in
opposition to the tide, obliged our intrepid
herring-fishermen to make several tacks,
which were all the more laborious, as there
were only two men, and they were without
many necessary articles of tackling.   All this,
however, was mere sport to men inured to
protracted fatigue, and accustomed, at the
fishing time, to brave all the dangers of a
dreadful sea.

On clearing the harbour, we came in
sight of that succession of islands which
skirt the Sound, and present a varied picture.
The isle of *Lismore* was at a very small
distance on our right; that of *Kerrera* in

the opposite quarter; and the peaks of Jura, known as the *Paps of Jura*, towered above the multitude of mountains with which the Hebrides are bristling. Lochaber,* which the largest vessels may navigate as high as Fort William, seemed to fly behind us. We discovered the isle of Mull; in the distance that of Skye; † and the ridge of Morven, so famous from the songs of Ossian, and so diversified in its landscapes, bordered the right side of the strait along which we sailed.

In passing near the end of the isle of Lismore, I observed, with the aid of my telescope, on a small neighbouring island, which was uninhabited, one of those monuments of rude stone known by the Hebridian appellation of *Cairn*.

This monument, of great antiquity, and erected in so desert a situation, naturally attracted my attention. I requested my companion to induce our boatmen to convey us thither; but as this small island, or rather large rock, is surrounded with breakers, they

* [Strictly Loch Linnhe; the name Lochaber is now restricted to a large district on the mainland, to the east of the upper part of Loch Linnhe.]

† [Skye is not visible from the Sound of Mull.]

answered that it would be impossible to land
on it except with the skiff, and that there
was even some risk to run.

As I did not understand a word of the
language of our conductors, but observed
one of our seamen preparing to enter the
wherry; I confidently followed him, and Mr
Macdonald did the same. The skiff was so
small and so shallow, that we had some
trouble to seat ourselves in it. The boatman
took the oars; Mr Macdonald steered the
helm, which consisted of the half of an oar,
and we pushed off.

Curiosity had here got the better of
prudence; we were borne along by the
current with the rapidity of an arrow, towards the small island, which is called
Fiart; * and it required all the address of our
boatman to land us safely. The isle is only
about half a mile in circumference, and rises
not more than twenty-five feet at most above
the sea. It may be regarded as a great rock,
of which the summit is flat and forms a small
plain. The action of the waves in a sea so

* [The word in the original is " Niort." The islet was one of
the group off the Rudha Fiart at the south end of Lismore.
The tidal currents here are among the most tumultuous in this
district of the west Highlands.]

battered by tempests everywhere lays the rock bare, and sweeps away the small quantity of soil which is formed there by time, so that nothing grows upon it except a few lichens and some scurvy-grass in the sheltered crannies. The rock is limestone, mixed with a little clay; it is of a blackish grey colour, and only forms a single mass, in which neither beds nor banks can be made out.

The kind of rustic pillar erected on this rock which had attracted our notice, is nine feet high, three feet broad, and two feet of average thickness. It consists of a grey granite, in which quartz and mica are predominant. The felspar is rather disposed in otreaks than in crystals, and though the texture of the stone is somewhat fissile, the base is hard and fresh in its fracture.

Though the shape of this kind of column is on the whole somewhat regular, one could not detect on it the slightest trace of workmanship; and one could only regard it as a natural longitudinal block, taken thus from the quarry, and erected on the highest point of the little isle, where it is sunk two feet in the earth, and kept upright by two solid stones, which give it a very stable foundation.*

Our boatman told Mr Macdonald that he had often seen this stone and that he knew it had been placed there by Ossian; and that in several other isles we should find much larger stones, erected by the same hands. For in the mountains of Scotland, and among the Hebrides, every thing that appears great, extraordinary, or marvellous, is always regarded as the work of Ossian.

Whatever may be the tradition respecting these ancient columns, this one evidently shows an intention to erect a simple but above all durable monument. The choice of stone plainly indicates this intention. It would doubtless have been more convenient to make use of the stone of the islet. But whether it was understood that this stone was much less durable than granite; or

---

* In some quarries of granite, and even at times in porphyric rocks, similar blocks are found, divided into parallelopipeds of various degrees of regularity and length. They are the effect merely of the contraction of the matter during the time of the aggregation of its molecules. Near the small town of Saint Siphorien-de-Lay, three leagues from Roane, there is a porphyric rock divided into large prisms, of which some are as remarkable for regularity as those of the largest and most perfect columns of basalt.

whether the use of iron was then unknown,
and there were no means of cutting out a
simple pillar from a calcareous rock which
is not divisible by beds; it is none the less
true that this rustic column of granite has
been transported hither; an undertaking
exceedingly difficult for men ignorant of
the mechanic arts.*

Though we spent but a very short time
in examining this stone, it was not without
great difficulty that we got back to the
vessel which the currents had caused to drift,
notwithstanding the efforts made to keep
near us. We were nearly an hour in
regaining her.

We continued our course through the
Sound of Mull, having always on our right
the granitic mountains of Morven. We
passed near the old castle of *Ardtornish*,
built upon a point which commands a view
of the whole Sound. The left side displayed to us the black volcanic rocks of
the isle of Mull. At length, after seven

* [The ice of the Glacial Period, by strewing blocks from
the Highland mountains all over the lower grounds, provided
the early inhabitants with great store of monoliths. It is
seldom necessary to suppose that these stones were brought
from a distance by human agency.]

hours and a half of navigation, we entered
the bay of Aros, the place of our dis-
embarkation in this island.

I hardly know what name to give to five
or six houses in a group, and seven or eight
others scattered around, the whole of which
taken together, bear the name of Aros.   It
is neither a town nor a village; it is rather
a hamlet; but it is most certainly inhabited
by very kind and very hospitable people.

The bay of Aros was anciently defended
by a strong castle, where the famous
Macdonald of the Isles resided; the ruins
of this fortress still remain; the building
was partly erected upon prisms of basalt.

We were received with the frankest
cordiality by an old gentleman, called
Campbell of Aros, living philosophically
in the modest habitation of his fathers;
a gothic building, standing on a brown
volcanic rock, totally destitute of every
kind of verdure, beaten by the hurricanes,
and yielding no other prospect than that
of a raging sea, fertile in shipwrecks.

Mr Campbell, wrapped in a large plaid
of variegated stuff, after the fashion of the
Hebridians, introduced us into his house,

and refreshed us with some port wine, sea biscuit, and preserve of myrtle-berries. His wife, who was not much younger than himself, and had perhaps never been out of her island, seemed much astonished to see strangers quit their native country, to visit a region so wild, and so difficult of access. Both of them pressed us very much to remain a few days with them; but as I was anxious to rejoin my fellow-travelling companions, who were waiting for me at Mr Maclean's, of Torloisk, we begged Mr Campbell to be so good as to procure us horses; which he had the goodness immediately to do. These horses were small, and had only a piece of rope for bridle; but they were good and accustomed to the bad roads of the country. We took leave of the old gentleman and his lady, and pursued our journey.

CHAPTER II

Journey from Aros to Torloisk.—Stay at Mr Maclean's.—Tale of what happened to my fellow-travellers during their passage to, and continuance on, the Isle of Staffa

THE miles of Scotland, particularly those of the isles, are nearly double the length of those in England. From our ignorance of this difference, we found that we had been much deceived when we were told, that from Aros to Torloisk was only eight miles. Imagining the computation to be by English miles, as we had set off at four in the evening, we conceived that we could easily ride that distance before dark.

I ought not to forget to mention that Mr Macdonald, who accompanied me from Oban with the intention of visiting the isle of Staffa, had no sooner reached Aros than he changed his dress. He had been in English regimentals, but upon arriving here he opened his portmanteau, and to my great surprise, in about an hour after, appeared in the complete habiliment of the inhabitants of the isles; plaid, jacket, kilt, feathered bonnet, buskin-hose, dirk in the waist-belt; nothing was forgotten. I could scarcely recognise him in this costume. He told me, that it was the garb of his fathers, that he never appeared in any other when in these islands, and that the wearing of it was a mark of attachment to his fellow-countrymen, with which they were much pleased.

We set out on our two little horses, with two persons to conduct us and bring them back, ignorant that our way lay across ravines, heaths, marshes and mountains, diffi-cult of access, and without any trace of a road.

Whilst day-light remained, we made good progress; our guides pushed forward with such speed as to outrun our horses, though these went at a good pace. Our two young Hebridians were well-made, light and inde-fatigable; they made nothing of streams, pools, bogs, or mountains; and I admired their courage, gaiety, and graceful figure. Their heads were decorated with a blue military bonnet, having a border of red,

green, and white, and surmounted with one
feather. They wore with grace a plaid,
having squares of different colours, fastened
to the right shoulder and folded over the left
arm, with a waist-coat and jacket of the
same stuff. Their thighs and legs were half
naked, but the latter were covered with a
coloured buskin, while a convenient kind
of shoes completed this Roman dress. A
poniard in the girdle gave them a military
air, and a stick in the hand served to help
them over the waters.

Their eagerness to be useful to us made
them doubly interesting. They never failed
to go before to show us the way, returning
however at intervals to caress and animate
the horses, or to ask if we had need of their
services.

They were delighted and proud to see a
man of distinction dressed like themselves; and
showed their satisfaction by coming up to Mr
Macdonald, and, with a smile on their faces,
telling him in their expressive language, that
they would follow him to the world's end.*

* Johnson also, in his journey to the Western Islands of
Scotland, praises two Highlanders whom he had hired as
guides on his way from Inverness.

Night was now coming on, and we were yet scarcely half way to the place of our destination. Our guides and our horses soon slackened their pace; the track grew detestable, and we were often obliged to alight, sometimes on the verge of marshes, and sometimes in the midst of heaths, from which we had a good deal of trouble to extricate ourselves. At length we completely lost our direction. The night was so dark that our horses fell down repeatedly, and our guides were much perplexed. After thus wandering for a long time without any certain course, we at last descried a light on rising ground, to which we directed our steps. It was the house of Torloisk, where we arrived at eleven o'clock at night, worn out with fatigue, vexation and uncertainty.

We soon discovered, as we entered, that we had now reached our place of destination.

"At Inverness," says he, "we procured three horses for ourselves and a servant, and one more for our baggage, which was no very heavy load.—We took two Highlanders to run beside us, partly to shew us the way, and partly to take back from the sea-side the horses, of which they were the owners. One of them was a man of great liveliness and activity, of whom his companion said, that he would tire any horse in Inverness. Both of them were civil and ready-handed. Civility seems part of the national character of Highlanders."

A domestic told our guides, that Mr Maclean
had not yet gone to bed, and that I had
been impatiently expected for several days.

We were shewn into a parlour, where I
found Mr Maclean, to whom I gave the
letter from the Duke of Argyll. He received me with the most obliging kindness,
and presented me to his wife, daughter, and
several other ladies and gentlemen, who were
engaged in a little musical concert.

Miss Maclean, a girl of a most charming
figure, played on a harpsichord some excellent
Italian music. Mr Macdonald had no need
of being introduced, his name was already
known, and his dress announced him. We
were instantly overwhelmed with civilities,
kindnesses, and delicate attentions, which
made all our troubles disappear. Everyone
was so prepossessing and so affable, that
from that moment we regarded ourselves as
in our own family.

How attractive is this country politeness,
seasoned with the expressions and gestures of
the most delicate feeling. We were here on
the true soil of hospitality; all the inhabitants
of this island, though the population amounts
to about six thousand souls, have only one

family-name, that of Maclean; they are distinguished only by their Christian names, or
by that of their residence; they are almost
all shepherds or fishermen.

They told us that my fellow-travellers
had sailed at five in the morning of the
same day, to visit the isle of Staffa; that
they would have waited with pleasure to
make the voyage in company with me;
but the season was already so far advanced,
and particularly the sea was so rough in
these parts, that they thought they ought to
profit by an interval of calm, which did not
promise to be of long continuance, so great
was their impatience to see that famous isle.

They had embarked with a friend of the
family and their own servants, in two small
boats. But they had scarcely gone four or
five leagues, before the weather suddenly
changed, and the sea became stormy. Mr
Maclean thought it so rough, that he was
afraid they would not have been able to
land on the isle of Staffa, surrounded as it is
with rocks, and that they had been obliged
to take refuge in the isle of *Iona* or *Icolmkill*,
which is fifteen miles from Staffa, and has
a small creek.

We expected that the sea would be a little calmer by the next day. We repaired therefore, at an early hour, along with Mr Maclean and his family, to the water-side, which was about three furlongs from the castle, to see whether the boatmen would venture to come for provisions; but the sea was still more dreadful, and totally impassable.

We now began to be very uneasy on their account. They were eight in number, including the domestics, and they had only one day's provision with them.

The evening arrived without any appearance of them; our anxiety was redoubled, and we passed a very unhappy night.

On the next day, which was Sunday, and the third day from their departure, I rose at four in the morning to inspect the weather. I discovered with pleasure that the wind was beginning to fall, and that the sea was not so high. We went, before noon, to take a walk on the shore; and at length, with the aid of a good glass, we descried them at a distance.

They arrived at one o'clock, to their own and our great satisfaction. They were so emaciated with fatigue, vexation and misery, were so much in want of food and rest, and so uneasy, that they entreated us not to disturb them with any questions until they were a little refreshed, and particularly relieved from a multitude of lice that tormented them most cruelly. " Fly! fly from us," said they, " we have brought some good specimens of mineralogy, but our collection of insects is numerous and horrible." We could not keep from laughing at this address, their gait, and the restless motion of every part of their body. They were instantly conducted to their apartments, where their first care was to clean themselves, to eat a little, and to take a few hours repose.

In the evening they returned to the parlour, where they were received with the greatest demonstrations of joy. Their appearance was now fresh and elegant; but that did not hinder us from asking, whether we might without risk come near to them. " We have cast off every thing," replied they, " and of all our evils there remains only the itch, respecting which, we can say nothing, as it is not yet developed."

They then recounted the circumstances of
their unfortunate passage. Notwithstanding
the fine appearance of the weather on the
day of their departure, scarcely had they
gone six miles, when a high wind worked
the sea into a terrible commotion. They
would have willingly put back, had not the
rocks which skirt the coast of Torloisk
made it equally dangerous of approach at
that moment ; moreover the currents and the
tide were also unfavourable. They were,
therefore, obliged to keep the offing, and
to brave the impetuosity of the billows,
driven sometimes in one, and sometimes in
another direction, and every instant in danger
of being swallowed up, but for the address
and experience of the boatmen, accustomed
from their infancy to this terrible sea.

Having, at length, after many struggles
and dangers, reached the isle of Staffa, they
found it still more difficult to effect a landing. By the assistance however of the
people of the isle, who, on seeing their distress, threw out some ropes to them, and by
watching a favourable wave, they reached
the shore without any other accident than
that of wetting themselves to the skin.

The coast however was too steep to admit
of hauling up the two boats, which were
obliged to put off again, and to take shelter
in the isle of Iona or Icolmkill, about fifteen
miles from Staffa.

Our friends, continuing their recital, informed us, that the only two families living
on this small island, received them with the
most affecting hospitality, and that the one
which thought itself in the most easy circumstances invited them to enter their hut,
where they were ushered into the midst of
six children, a woman, a cow, a pig, a dog,
and some fowls.

There was laid out for them a remnant of
oaten straw which had been used to litter the
cow for several days before. This served as
their seat, table, and bed. A fire of bad
peat, or rather ill-dried sod, lighted in the
middle of the cabin, smoked them, at the
same time that it dried their clothes and
served to cook, in an indifferent manner,
some potatoes, which, with a little milk, were
the only articles of food the place afforded,
and these in very small quantities. The
provisions which they had brought with
them were consumed at a single meal.

The sea broke upon the island with such impetuosity, and rushed with so much turmoil into the caves with which it is pierced, that the hut shook, and our adventurers could not shut an eye.

Next day it rained incessantly until noon. The sea, far from calming down, raged with still greater fury; so that the boatmen could not venture to carry any supply of provisions from the isle of Iona.

In the afternoon, the rain having ceased, the captives surveyed the island and visited Fingal's Cave. William Thornton took some views of it with much care; and they made a collection of the most curious stones, among which were some fine zeolites.

In the evening they had the same reception, the same supper, and the same bed. A new incident, however, occurred: The master of the cottage, his wife, and children lived in such a horrid state of filth that the place was as full of vermin as of wretchedness. Detachments of lice approached from all sides to pay their respects to the new lodgers, who were soon infected with them. This was the most cruel of their torments, and formed the object of an occu-

pation which did not allow them a moment's
respite.

On the third day, the sea was somewhat
calmer. Their distress was at its height.
They walked repeatedly round the island,
and ascended the highest part of it to look
for the approach of the boats, which at length
made their appearance, and came to deliver
our poor friends from their afflicting captivity.
After recompensing their hosts for their kind
offices and hospitable attention, they took
leave of them to return to Torloisk ; where
we had the happiness of welcoming them
with all the ardour of friendship, congratulat-
ing them that they had got off so cheaply as
to escape with only a few days of abstinence.
Finding them all safe, it was not without
laughter, that we heard them relate their
misfortunes, and particularly the diverting
episode of the lice.

Their account brought to my remembrance,
at the moment, a similar adventure which
happened in the same isle, and probably in
the same house, to Sir Joseph Banks, who
set out from London in the year 1772,
on a voyage to Iceland, in company with
Solander, James Lind, Gore, Walden, and

Troil; and, in passing, paid a visit to the fine cave of Fingal, which he was the first to make known.

On their arrival at Staffa, they erected a tent, to pass the night under it; but the only inhabitant then on the island pressed Sir Joseph so strongly to come and sleep in his hut, that he out of complaisance consented, and left his companions under the tent.

On leaving the hut next morning, he discovered that he was completely covered with lice. He mentioned the circumstance to his host in terms of mild reproach. But the latter, who was touched to the quick, perked himself up, and assuming a tone of consequence, retorted haughtily and harshly that it was Sir Joseph himself who had imported the lice into his island, and adding, that he had better have left them behind him in England.

The detail of the adventures of my poor friends, did not much encourage me to attempt the same voyage. Mr Maclean also did not cease on his side to impress me with the inconstancy of this sea, the dangers in landing on the island, the advanced period of the season, and his fear that even suppos-

ing we could seize a favourable moment
to get there, we might not find it equally
easy to get back, and might be obliged to
remain there, not merely several days, like
our friends, but perhaps for several months.

"I am now old," said Mr Maclean; "I
have made several voyages to India, and I
am accustomed to the sea; but every time
that, from complaisance to any of the persons
recommended to my attention, I have accompanied them to the isle of Staffa, which
lies, so to speak at my door, I have had
occasion to repent it. I have made this
voyage six times in my life, on the most
favourable days and with excellent fishermen,
and I have always run into some kind of
danger, either in going or returning. The
landing is especially terrible even with the
smallest boats, so steep is the coast and so
furious is the sea around this island.*

All this, I repeat, was not very en-
couraging, especially to one, who, like me,
is almost always sick on the water; but
curiosity overcame fear and prudence. What,

* "Here," says Mr Pennant, "Æolus may be said to
make his residence, and be ever employed in fabricating blasts,
squalls, and hurricanes."

said I continually to myself, shall I have come, so to say, to the very entrance of this renowned cave, and from such a distance too, without landing at it? What! I should thus forego the opportunity of obtaining new information and instructive facts, in a part of natural history which so much interests me as that of ancient volcanos, and I should not do what my fellow-travellers have done; I should not run into the same dangers. All these reflections irrevocably fixed my determination; and I resolved to set out at sunrise next morning, if the sea should be anywise passable.

I instantly had a boat engaged. Mr Macdonald said that he would accompany me; and my intrepid friend, William Thornton, scarcely yet recovered from his fatigues, in spite of all the dangers he had already met with, told me that he was also ready to begin again, and would come with me. This young American had so strong a desire for instruction, and so keen a taste for natural history, that nothing could damp his ardour.

CHAPTER III

Voyage to Staffa

NEXT morning at four o'clock, one of our boatmen came to inform us that the weather appeared to be promising, and that it was probable we should have a fine day. Having made the requisite arrangements the preceding evening, we were soon ready, and reached the beach before sunrise.

Our rowers were four young and bold Hebridians, who appeared to undertake this short voyage with pleasure; for they love every thing that reminds them of Ossian, and they seemed to regard it as a happiness and honour to conduct strangers to Fingal's Cave. We besides, allotted them a quantity of refreshments, of which, to be prepared against whatever might happen, we laid in an abundant provision.

The boat was very small and incapable of carrying a sail. Our four seamen seated themselves on their benches; Mr Macdonald

took the helm, William Thornton and myself sat down on a bundle of sea-weed,
and we put off under the auspices of the
genius that presides over the natural sciences,
to whom we made a short invocation.

It took us hardly an hour and a half to
double the point of the isle of Ulva, opposite
to that of Mull, near Torloisk, whence we
had sailed. We then entered the open sea,
and soon found, that in these parts, the
ancient and majestic ocean has no need of
being agitated by the northern blasts to rock
up and down and to toss itself into big waves.

Continuing our course, we had a view of
the volcanic isles of Bac Beg, and the Dutch-
man's Cap, with those of Lunga, Skye,
Gometra, Iona, &c.

We could not have wished for a more
agreeable passage at so advanced a season.
Our seamen, making Mr Macdonald their
interpreter, assured us, that it was one of
those extraordinarily fine days in this country
which seldom come twice in the year. So
to show their pleasure, they began to chaunt
in chorus the songs of Ossian. There is
nobody in these islands, from the oldest to
the youngest, who does not know by heart,

long passages or hymns of that ancient and
celebrated bard.

The songs began and continued a long
time. They consisted of monotonous re-
citatives, ending in choruses equally monotonous. A sort of dignity, mingled with
plaintive and melancholy tones, was the chief
characteristic of these songs. The oars,
which always kept time with the singing,
tended to make the monotony more complete.
I became drowsy, and soon fell sound asleep.

I know not how long this lasted; but I
was awakened by the movement and noise of
the seamen, and was told that we were now
close upon the isle of Staffa, and near some
reefs, which required new manœuvres. Here
I had an opportunity of witnessing, not
without dread, the address and intrepidity of
our sailors, who knew how to seize the
favourable moments to avoid being dashed
to pieces, and to choose the propitious waves
which afford a safe passage over the reefs
that make this landing so dangerous.

Two of the inhabitants of the island soon
ran towards us, and from the top of their
rock threw down some ropes to us; with this
assistance and the aid of a large wave selected

on purpose, we disembarked amidst a cloud
of foam.

These two men led us, and our small
crew, to a level spot on the top of the isle,
where stood two houses, or rather huts, constructed of large blocks of lava and broken
prisms of basalt, and covered over with green
sods, getting no light except from the door,
which was only three feet high, and from
the chimney, which consisted of a pyramidal
opening in the middle of the hut.

The women and children of the two
families did not fail to come to us, and
invited us to their habitations : but being
already informed of their excessive dirt, we
were inflexible; and preferred on good
ground to receive their civilities and their
compliments in the open air.

Finding, that it was impossible to prevail
with us by the most friendly gestures, they
resolved to do us the honours on the small
esplanade in front of their dwellings.

The men, women, and children, with
much gravity, first formed themselves in a
large circle in which they placed us and
our seamen.   Then one of the women,
disgustingly ugly and dirty, brought out

a large wooden bowl filled with milk, with
which she placed herself in the centre of the
circle. She viewed us all round with attention, and immediately came up to me, and
pronouncing some words, presented the
bowl with a sort of courtesy. I held out
my hands to receive it; but she drank
some of it, before she gave it to me. I
followed her example, and passed the vessel
to my neighbour William Thornton, he
gave it to Mr Macdonald; and so on from
hand to hand, or, more properly, from
mouth to mouth, till every person had
tasted of it. Having made our acknowledg-
ments for this kindness, they immediately
appointed two guides to accompany us
to Fingal's Cave and all the remarkable
places of this small isle. We ate a morsel
of bread, to take off the edge of our appetite
during the walk; as it was agreed upon,
that in order to lose as little as possible of
so favourable a day, we should postpone
taking our repast till we were seated in the
boat on our way back. This allowed us
sufficient time to see with ease all the objects
of curiosity in the island, and particularly
to direct our attention to that remarkable

cave which we had come so far to see, and
which we congratulated ourselves on being
able to study on one of the finest days of
the year.

We went to work, therefore, without losing a moment of time. I soon arrived at
the entrance of this marvellous monument,
which, according to an ancient, but fabulous
tradition, was the ancient palace of the
father of Ossian. I was obliged to put off
my shoes in order not to slip in the depth
of the cave into which the sea rushes with
great noise, and where one cannot advance
but by walking with the utmost precaution
along a sort of cornice on the right side,
about fifteen feet above the surface of the
water, and formed of a multitude of vertical
basaltic columns, on the broken tops of
which one must dexterously place one's feet,
at the risk of falling into the sea, which
runs in to the furthest recess.

Attention is so much the more necessary
here, as the cliff along the top of which one
has to tread is entirely perpendicular, and
the ledge is in some places at most two
feet wide, formed of unequal prisms, very
slippery, and constantly wet with the foam

of the waves and the dripping moisture from above. The light, which comes only from the grand entrance and diminishes gradually as one advances inwards, serves to make the path still more difficult.

I ceased not to view, to review, and to study this superb monument of nature, which in regard to its form bears so strong a resemblance to a work of art, though art can certainly claim no share in it. I took all the measurements of it, with the assistance of Mr Macdonald, who was of the greatest use to me. I wished to observe the most scrupulous exactness in that operation, and he perfectly seconded my intentions.

During this time, my indefatigable friend, William Thornton, took a drawing of the cave, which could be seen in a true point of view from the sea only. [Plate IV.] This task was neither agreeable nor free from danger ; for it required all the address of our seamen to keep for a few moments in front of the entrance, amidst the whirlpools and waves of a sea which seemed eager to engulf the frail skiff. It was necessary continually to return to the same point, and to give rest at

intervals to my dear Thornton, who grew
sick with the tossing.

Our ardour and perseverance were not to
be shaken, and nothing could distract our
attention. We only looked abroad, from
time to time, to see if the sea would treat
us as favourably during the rest of the day.
After having noted down all the particulars
respecting Fingal's Cave, after having
sketched the objects that most interested
us, and after having taken the measurements which I was glad to obtain, I went
on to examine the other parts of the island;
and I made a collection of different lavas,
zeolites, and other stones, which might
serve to illustrate the natural history of
the place.

I saw with some uneasiness that the sun
was now about to leave us, and that we
must make up our minds to tear ourselves
away from a place which presented scenes
so striking and volcanic phenomena so remarkable. But the weather might change
from one moment to another and we had
a long passage to make. So we left the
island, re-embarking at half past four in
the afternoon. We took some refreshment

on the way, for we were famished. Our
indefatigable Hebridians, who felt neither
our curiosity nor our tastes, except for
Fingal's Cave, for which they have a sacred
veneration, had made a hearty meal on the
island, and had thus diminished the weight
of our stores, during the time that we
were busy seeing and observing everything.
They were pleased, and set themselves to
row us along with a spirit and vigour, which
were at once a proof of their strength and
of their habitual capacity for toil. They
were so delighted to be able to carry us
back safe and sound, on so lovely a day,
and on so smooth a sea, that they became
enthusiastic, and never stopped their singing.
At nine o'clock in the evening we were back
at Torloisk, where the good Mr Maclean,
his family, and our friends, were impatiently
expecting us.

I employed myself during several days in
digesting my observations on the isle of
Staffa; and for the sake of greater method
and perspicuity in what I have to say, I
thought proper to adopt the following order.
The reader will be pleased to recollect that
this description is principally intended for

such as pursue the study of the natural history of stones and minerals. If it be considered as rather tedious by those who are not attached to that study, it will be easy for them to pass on to other subjects.*

* [The author criticises (*note* p. 45) the representation given in one of the plates accompanying the description of Staffa by Sir Joseph Banks, but his own plates are far from accurate. The view of the interior of Fingal's Cave gives an erroneous impression of architectural regularity and depth. In the general view of Staffa, besides the defective perspective, the title reverses the position from which the sketch was made. The view represents the south-eastern, not the north-western front of the island.]

View of the Isle of Staffa, from the North West.) with the entrances of the Cave of Fingal, and the Cave of the Cormorants.

CHAPTER IV

*Description and Natural History of the
Isle of Staffa.—General Views*

THE isle of Staffa is situated in the fifty-
seventh degree of north latitude, and
fifteen miles west of the island of Mull. Its
form is oblong and irregular. Its coast is
everywhere a succession of cliffs, surrounded
with the most superb basaltic causeways,
and hollowed into different caves, such as
those of Fingal and the Cormorant. The
isle is accessible only by a small opening or
entrance, where the escarpment is less
sheer and sinks into a slope, but which can
admit only a small boat, and that in calm
weather, for if there be the least wind,
landing becomes dangerous, and the boat is
obliged to seek shelter at the island of Iona.

The total circumference of Staffa is little
more than two miles. The highest part is
above Fingal's Cave, where it is one hundred
and fourteen feet above the level of the sea
in ordinary tides.

The whole framework of this grand
volcanic rock is laid bare; the waves and
currents seem to batter and mine it from
every side. Only on the highest part is
there a flat piece of ground covered with
a poor dry turf, alongside of which lies a
corner of ground newly broken up, where
a little oats and a few potatoes are raised.
There is also a small pasturage and a scanty
spring, which would be soon dried up, were
not the climate so rainy.

Not a tree, not a bush is to be seen, and
for firing, the inhabitants are obliged to
make use of a bad turf, which they lift
in the summer season in order to dry it.
It is not peat; since it consists simply of
the fibrous roots of common grasses intermixed with earth. Nothing worse in the
way of fuel could be used; but here
necessity reigns with absolute sway.

The whole of the isle belongs to Colonel
Charles Campbell, of Campbeltown, in Can-
tyre. It is let at the rent of twelve pounds
sterling; on account, probably of its fishery,
for its territorial value ought to be considered
as nothing.

The total population, at the time when I

visited it, consisted only of two families, who lived apart in two huts, constructed of unhewn blocks of basalt, roofed over with sods, and who amounted, men, women, and children, to the number of sixteen.* Belonging to these, there were eight cows, one bull, twelve sheep, two horses, one pig, two dogs, eight hens, and one cock.

Buchanan has slightly mentioned the isle of Staffa and its remarkable columns. But Sir Joseph Banks, President of the Royal Society of London, is the first who examined this grand and astonishing object of natural history with the eye of an observer. He has given it celebrity by his description of it, which was published in the Tour to the Hebrides, by Thomas Pennant, accompanied with plates.

Mr Troil, Bishop of Linkœping, one of Sir Joseph Banks's fellow-travellers, has

* At the time when Sir Joseph Banks, in 1772, visited this island, along with several naturalists, of whom Mr Troil was one, it belonged to Mr Lauchlan Macquarrie and it had only a single inhabitant.

"There is only one hut," says Mr Troil, "which is occupied by a peasant, who attends some cattle that pasture there. To testify his joy for our arrival, he sung all night over in the Erse language, which we did not understand. He regaled us with fish and with milk."—*Letters on Iceland*, by U. von Troil, Bishop of Linkœping.

given a description of the same isle, and
of Fingal's Cave in a learned and curious
work upon Iceland.* But as these two
travellers have principally attended to the
picturesque scenes, without entering into
those details which are more particularly interesting to naturalists, I conceived that I
might give some satisfaction by pursuing the
latter track.

Of the Cave of Fingal, or An-uamh-binn

This superb monument of a grand subterranean conflagration, the date of which has
been lost in the lapse of ages, presents an
appearance of order and regularity so
astonishing, that it is difficult for the coldest
observer, and one the least sensible to the
phenomena which relate to the revolutions

* This work, written in the Swedish language, has been
translated into French by M. De Lindholm, and printed at
Paris by Didot 1781, in 1 vol. 8vo, with plates. One could
wish that the translator, to whom the Sciences are indebted for
rendering that excellent book into our language, had been more
acquainted with natural history; his notes would then have had
more interest, and contained fewer errors. [Uno von Troil
(1746-1803) was afterwards (1787) archbishop of Upsala.
His Swedish work, " Bref rörande en Resa till Island " was
translated also into German and English. The English
version will be found in Pinkerton's "Voyages," vol. i.
(1808) p. 621.]

of the globe, not to be singularly surprised by this kind of natural palace which seems a veritable marvel.

To shelter myself from all criticism of my feelings, while contemplating the most extraordinary of all known caverns, I shall borrow the expressions of him who first described it. Those who are acquainted with the character of this illustrious savant, will not accuse him of being liable to be hurried away by the fire of a too ardent imagination; but the sensations which he felt at the view of this magnificent scene were such, that it was impossible for him to resist a justifiable enthusiasm.

" The impatience which everybody felt to see the wonders we had heard so largely described, prevented our morning's rest; every one was up and in motion before the break of day, and with the first light arrived at the S.W. part of the island, the seat of the most remarkable pillars; where we no sooner arrived than we were struck with a scene of magnificence which exceeded our expectation, though formed as we thought upon the most sanguine foundations. The whole of that end of the island, supported by ranges of natural pillars, mostly above fifty feet high,

standing in natural colonnades, according as
the bays or points of land formed themselves, upon a firm basis of solid unformed
rock. . . . In a short time we arrived at
the mouth of the cave, the most magnificent,
I suppose, that has ever been described by
travellers.

" The mind can hardly form an idea more
magnificent than such a space, supported on
each side by ranges of columns, and roofed
by the bottoms of those which have been
broke off in order to form it ; between the
angles of which a yellow stalagmitic matter
has exuded, which serves to define the angles
precisely, and at the same time vary the
colour with a great deal of elegance ; and
to render it still more agreeable, the whole
is lighted from without ; so that the farthest
extremity is very plainly seen from without,
and the air within being agitated by the flux
and reflux of the tides, is perfectly dry and
wholesome, free entirely from the damp
vapours with which natural caverns in
general abound." *

* " A Tour in Scotland and Voyage to the Hebrides,"
MDCCLXXII. By Thomas Pennant, 2nd Edit. 1776, Part
I. p. 300.

Let us also, for a moment, listen to Mr Troil upon the same subject.

" How splendid," says this prelate, " do the porticos of the ancients appear in our eyes, from the magnificence displayed in the descriptions we have received of them, and with what admiration do we look even on the colonnades of our modern edifices ! But when we behold the cave of Fingal, formed by nature, in the isle of Staffa, it is no longer possible to make a comparison, and we are forced to acknowledge that this piece of architecture, executed by nature, far surpasses that of the colonnade of the Louvre, that of St Peter at Rome, and even what remains to us of Palmyra and Prootum, and all that the genius, the taste, and the luxury of the Greeks could invent." *

Such was the impression made by Fingal's Cave on Sir Joseph Banks, and on the Bishop of Linkœping. I have seen many ancient volcanoes; I have described and made known some superb basaltic causeways and fine caverns in the midst of lavas ; but I have never found any thing which comes near this one, or can be compared to it, for

* " Lettres sur l'Islande," p. 376.

the admirable regularity of the columns, the height of the arch, the situation, the forms, the gracefulness of this production of nature, and its resemblance to the master-pieces of art: though art has had no place here. It is therefore not at all surprising that tradition should have made it the abode of a hero.

The mouth of this fine monument of nature is thirty-five feet wide and fifty-six feet high, and the cavern has a length of a hundred and forty feet.

The upright columns which compose the façade, are of the most perfect regularity. Their height to the beginning of the vault is forty-five feet.

The concavity of the roof is composed of two unequal segments of a circle, which form a sort of natural pediment.

The mass which crowns, or rather forms, the roof, is twenty feet thick in its thinnest part. It is composed of small prisms, more or less regular, inclining in all directions, closely united and cemented underneath and in the joints with a yellowish white calcareous material, and some zeolitic infiltrations, which give this fine ceiling the appearance of mosaic work.

The sea reaches to the very end of the cave. It is fifteen feet deep at the mouth ; and its waves, incessantly agitated, beat with great noise against the bottom and walls of the cavern, and every where break into foam. The light also penetrates through its whole length, diminishing gradually inwards, and exhibiting the most wonderful effects from the varieties of light and shade.

The right side of the entrance presents, on its outer part, a tolerably large amphitheatre, formed of different rows of large truncated prisms, on the top of which one can easily walk. Many of these prisms are jointed, that is, concave on the one side, and convex on the other; others are divided by simple transverse sutures.*

The prisms, of a black basalt, very pure and hard, are from one to three feet in diameter. Their forms are triangular, tetrahedral, pentagonal, and hexagonal; some have seven or eight sides. I noticed several

* Sir Joseph Bank's draughtsman, very good and accurate in other respects, has substituted, probably to give greater effect to the cave, large masses of stone irregularly piled on each other, on the right side of the kind of amphitheatre, which serves as a basis to that part of the grotto. But there is in reality nothing there except columns.

large prisms on the truncated tops of which
one can distinctly trace the outlines of a
number of smaller prisms; that is, these
prisms are formed of a basalt, which has a
tendency to subdivide into prisms in such a
manner that a large prism is composed of
outlines of several smaller ones.   I had before
observed the same phenomenon among the
prismatic basalt of the Vivarais.

The cave can be entered only on the right
side by following the platform which I
have mentioned.   But the track grows very
narrow and becomes increasingly difficult
as one advances; for this sort of interior
gallery, raised about fifteen feet above the
level of the sea, is formed entirely of truncated prisms placed vertically and of a
greater or less height, between which considerable address is necessary to choose one's
steps, the passages being sometimes so strait
and so slippery, owing to the drip from
roof and walls, that I took the very prudent
precaution, suggested by our two guides, to
proceed barefooted, and to take advantage of
their assistance, especially in one place, where
there was room only to plant a single foot,
whilst I clung with the right hand to a large

prism for support; at the same time that
with the left hand I caught hold of the hand
of one of the guides. This difficult operation took place at the darkest part of the
cave; where the half of one's body hung
over an abyss, in which the sea is so agitated
as to send up a cloud of spray.

I was desirous of penetrating to the very
end, and I succeeded though not without
difficulty and danger. I sometimes felt
distracted from the observations which I
was glad to make by the thought of how
I should get back again.

As one comes near to the end of the cave,
the risky balcony, on which one has walked,
expands into a large sloping space composed
of thousands of broken vertical columns.
The extremity of the cavern is then found
to end off in a wall of columns of unequal
height, resembling the front of an organ.

It is worthy of remark, that at the time
when Mr Troil visited the cave, the sea, by
one of those uncommon chances which do
not happen once in ten years, was so calm
that it allowed him to enter in a boat.

" At the very end of the cave," says Mr
Troil, " and a little below the surface of the

water there is a kind of recess which sends forth a very agreeable noise every time that the water rushes into it." *

As the sea was far from being completely still, when I visited the cave, I heard a noise of a very different kind every time that the waves, in rapid succession, broke against the end. This sound resembled that which is produced when a large hard body strikes heavily and with force against another hard body in a subterranean cavernous place. The shock was such as to be heard at some distance, and the whole cavern shook with it. Being close to the place whence the sound issued, and where the water is not so deep upon the retreat of the wave, I endeavoured to discover the cause of this terrifying shock, and was not long in observing that, a little below the base which supports the organ-fronted colonnade, there is an aperture which forms the outlet of a cavity, perhaps even a small cave. It was impossible to penetrate into this cavity, but it may be presumed that the tremendous noise is caused by a loose block of stone when driven forward with extreme violence by the surge against

* " Lettre sur l'Islande," p. 379.

the sides of the recess. On the other hand it can be noticed by the boiling motion of the water in this place, that there are several other small passages, through which the water escapes, after rushing into the principal aperture in a mass. It is, therefore, possible, when the sea is not sufficiently agitated to set the emprisoned block in motion, that the air, strongly compressed by the weight of the water, which is in incessant fluctuation, should, on rushing out by the small lateral passages, produce a particular strange sound. It might then be truly regarded as an organ made by the hand of Nature; this explanation would fully account for the ancient and real name of this cave in the Erse language being, *the melodious cave*.*

* Sir Joseph Banks is the first who gave the cave of Staffa the name of Fingal's Cave. I made the most minute enquiries of several persons well skilled in the Erse, Gaelic, or Celtic language, and especially of Mr Maclean of Torloisk and Mr Macdonald of Skye, to know what relation this cave could have to the father of Ossian. And these gentlemen, as well as others, assured me, that the mistake was owing to the name being equivocal. The following is their explanation : The true name of the cave is *an-ua-vine*. *An*, the ; *ua*, grotto, cave, cavern ; *vine*, melodious. The name of Fingal in the same language is spelled and pronounced *Fion* in the nominative. But the Erse nouns are declinable, and the genitive of Fingal is *Fine* ; so that if one wished to mention the cave of Fingal in the Erse language, he would write

Sir Joseph Banks, in the description which he has given us of the cave of Staffa, says that " between the angles a yellow stalagmitic matter has exuded, which seemed to define the angles precisely." That is true, but the learned writer has not told us the nature of this yellowish matter.

Mr Troil mentions it also, saying, that " the colour of the columns is a dark grey; but the joints are filled with a crust of stalactitic quartz, which distinctly marks the separation of the columns, and which, by the variety of its tints, has the most agreeable effect on the eye." There is certainly an error here with regard to the substance.

an-ua-fine. Thus between the Erse *vine* [binn] melodious, and the genitive of Fingal *fine*, there is no other difference than the change of the letter *v* into *f*; and some person not very well versed in the Erse language, might have translated to Sir Joseph Banks the words *an-ua-vine* [an-uamh-binn] by the *cave of Fingal*, whilst the true and literal interpretation is, the *melodious cave*. In this case, the observation of Mr Troil, on the agreeable sound which he heard issuing from the bottom of the cave when the water rushed in, is valuable, and comes in support of the true interpretation. [There can be little doubt that this etymological explanation is correct. The volume of air driven into the internal cavity by the impact of the water is under a pressure sometimes of several tons to the square inch of surface. When this pressure is suddenly relaxed by the sinking of the wave, the imprisoned air at once rushes out and, in favourable chinks and passages, gives rise to musical sounds as when a trumpet is blown.]

On breaking off several pieces of it, which
it is not very easy to do, owing to the height
of the vault, I found it to be nothing but a
calcareous substance coloured by the decomposition of the iron of the lava, and intermixed with a little argillaceous earth. This
kind of stalactite has also very little cohesion,
and is in general earthy. In several of the
prisms I found some kernels of zeolite, but in
very small quantity. I also broke off from
between two prisms, which were so far apart
as to allow the hand to be thrust between
them, an incrustation in which the white
and transparent zeolite appeared in very
perfect small cubical crystals, several of which
were coloured red by the ferruginous lime
arising from the decomposition of the lava.
But I repeat, that zeolite is very rare in
this cave, and having myself broken off
the few specimens that I was able to see,
I doubt whether those who may visit the
place after me will find much of it.

*Measurements and dimensions of Fingal's
Cave*

Breadth of the entrance, taken at the mouth
and at the level of the sea, thirty-five feet.

Height from the level of the sea to the
  beginning of the curve of the vault, fifty-
  six feet.

Depth of the sea, opposite the entrance, and
  twelve feet distant from it, at noon of the
  27th of September, 1784, fifteen feet.

Thickness of the roof measured outside from
  the beginning of the arch to its highest
  part, twenty feet.

Length of the cave from the entrance to the
  end, one hundred and forty feet.

Height of the tallest columns on the right
  side of the entrance, forty-five feet.

Depth of the sea in the interior part of the
  cave, ten feet nine inches; in some places
  eight feet, and towards the end somewhat
  less.*

I have given a description of the largest
cave, as the most remarkable. But there
is a second on the way to the northern part

---

* All these measurements were taken with great exactness
with a piece of thread-tape, painted and waxed, divided into
French toises, feet and inches, and rolling up into a leather
case. This instrument, which I had made for me in London,
afforded a measure of 100 feet. If I, therefore, differ in the
least from the dimensions taken by Sir Joseph Banks, attention
must be had to the difference of the English foot. This
naturalist, besides, used a fishing line, which, stretching more
or less with the wet, never gives the measures so correctly.

of the island, in the midst of a fine colonnade.
Its name in the Erse language is *Ua-na-Scarve* [Cormorant's Cave]. It is, however,
less interesting than the first, and was, besides,
inaccessible at the time I visited the island.
There is also at the south end, and not far
from the landing place, a small cave, in the
compact lava, surmounted with a set of prisms
which, as observed by Sir Joseph Banks,
exactly resemble the bottom of a vessel with
her sides exposed to view. The curvature
of the prisms makes this similarity striking.

More than one half the circumference of
the isle is occupied by very fine colonnades,
which are completely exposed on the side
next the sea. They rest, in general, on a
current of gravelly lava, which serves for their
base and support; and they follow the direction more or less inclined, more or less horizontal of this current. All these prismatic
causeways are covered with an enormous
stream of lava, more or less compact, and
tending more or less to a prismatic form.
The summit of this crowning sheet is covered
over with a little vegetable soil formed by decomposed lava, and with some thin common
grasses which grow there.

Above one half the island, therefore, is supported on columns more or less vertical; all the rest consists entirely of lavas more or less compact, more or less decomposed, more or less intermixed with fragments of other lavas, zeolitic infiltrations, calcareous streaks, and chalcedonic oozings, which have in some places penetrated even the substance of the zeolite.

One of the causeways to the north of the great cave merits the attention of the naturalist for the disposition, the number, the regularity, and height of the prisms, which are more than forty-eight feet high, and are placed vertically like the pipes of an organ. This magnificent colonnade is overspread with a current of compact lava, more than fifty feet thick, and composed of innumerable small prisms which diverge in all directions. It rests upon a current of black gravelly lava, nine feet thick, the paste of which is an intermixture of different other lavas divided into small irregular fragments, and united by a natural cement, composed of calcareous earth, zeolite, and a chalcedonic substance. Every thing leads me to regard this current as the result of a volcanic

View of the Basaltic Island of Boo–Schala: Adjoining to Staffa.

eruption, in which the water entering into co-operation with fire, has imbedded all these materials. A part of this current of lava lies under the sea.*

Not to extend this description too far, I have only to say a few words respecting what is improperly called the isle of *Boo-sha-la* [*Buachaill*]; † I say improperly, because the name of isle can never be given to what evidently forms an appendage of the principal isle. [Plate VI.]

Boo-sha-la lies at a small distance from the great cave, and is separated from the main island only by a channel a few fathoms wide; but the junction of this little volcanic hill with Staffa may be easily traced in the sea. Boo-sha-la itself seems to be divided into two parts at high tides. It is composed of a number of eminences of prismatic basalt of a very regular kind, grouped together in bundles in some places, curved in others, and

---

* [This "current of lava" is a bed of volcanic tuff and conglomerate due to the eruption of ashes and stones during intervals between the successive outflows of basalt.]

† ["Buachaill" is the Gaelic word for a herdsman and is often applied in the Highlands to solitary rocks or mountains which seem to keep guard over the surrounding ground, like a shepherd over his flock : thus the conspicuous Buachaill-Etive at the head of Glencoe is the "Herdsman of Etive."]

sometimes disposed in the manner of steps, which form a passable, though steep staircase. By the side of this the columns are vertical, and form by their union and their different degrees of elevation, a regular conical peak, which is entirely an assemblage of prisms. This remarkable structure is not due to displacement, nor to the slipping down of large masses. It seems rather to be the effect of a more or less gradual cooling; and the material in shrinking appears to have undergone fantastic modifications and accidents, comparable with those which may be observed in crystallizations on a large scale; though I am far from considering the prismatic lava as the result of crystallization. On the contrary, I reject that opinion; and the comparison which I use here is only for the purpose of making myself more intelligible, and has no relation but to the accidental varieties and different dispositions of the forms.

Mr Thomas Pennant has published, in his " Tour to the Hebrides," two engravings of Boo-sha-la, taken from the very correct drawings of Mr Banks.*

* *Op. cit.* Pl. xxix., xxxi.

It remains only that I should publish at
present a note of the lithological productions
of the isle of Staffa.

Mineralogy of the Isle of Staffa

1. Triangular basaltic prisms, which are
here, as elsewhere, rare.

2. Quadrangular; equally rare.

3. Pentagonal: } These are the most
4. Hexagonal : } common forms.

5. Heptagonal, which are found here.

6. Octagonal, of large size, sometimes
four feet in diameter, exhibiting in their
cross-section indications of other smaller
prisms.

7. Articulated prisms, that is, whose
sections are concave on one side and convex
on the other.

8. Prisms cut straight without articula-
tions ; some have eight, ten, and even twelve
sections.

9. Prisms in one piece ; these may be
seen twelve, fifteen, twenty, and even forty
feet high.

10. Prisms curved in the arc of a
circle.

11. Black gravelly compact lava, which easily separates into irregular pieces.

12. Black porous lava. The extinct volcano of the isle of Staffa, has been exposed for so many ages to the action of a sea full of currents, and subject to so many tempests, that one may well say that nothing but the skeleton of a volcanic isle, can here be seen which was once much more considerable; the sea attacking the island from every side has carried away or destroyed every thing that it could overcome. We should not, therefore, be astonished to find here neither the remains of a crater, nor scoriæ, nor light lavas. The same thing has happened to several other extinct volcanos which the sea has ultimately abandoned, after a perhaps incalculable lapse of time. On examining, however, with attention the materials that compose the currents of lava, on which many of the prismatic causeways of the isle rest, one discovers fragments of a black porous lava. These lavas being mixed among the debris of other lavas, compact, pulverulent or gravelly, compressed by the enormous weight of the overlying masses, and united by a cement partly calcareous

and partly zeolitic, are thus more protected from the action of the waves.

13. White radiated zeolite, incrusted in a basaltic lava. The same zeolite in a black lava, much less hard, in round, oval or irregular pieces, and in diverging needles. On the outer surface of these oval pieces, projecting crystals of cubical zeolite may sometimes be observed.

14. White radiated chalcedonic zeolite. I obtained from one of the deposits of muddy lava, on which the greater part of the prismatic causeways of Staffa rest, spherical kernels of zeolite with diverging rays, united to the number of three or four in one group. Several of these small balls were completely solvable in nitric acid, with which they formed a jelly; whilst several others adhering to these, but semi-transparent and of a greasy lustre, were impervious to the acid, and even gave sparks with steel. But on calcining and reducing the latter to powder, and digesting them in a glass vessel with nitric acid in a sand bath, the acid dissolves the zeolites and forms a jelly with it, whilst the chalcedonic particles, remaining untouched, are precipitated to

the bottom. I found some of these small balls of the size of a gall-nut, the one half of which had been penetrated by a chalcedonic milky juice, and the other by a quartzose juice extremely crystalline, and as transparent as the purest rock-crystal.

15. Cubical white zeolite. There were some of the most superb pieces of cubical zeolite in Staffa; but, in our visit to that isle, we took away all that were most interesting. Before us, Dr Thompson had also made at Staffa a very interesting collection of zeolites, and among others, a number of large cubic crystals clustered upon a black compact lava. This specimen, the most considerable and the most perfect of its kind, may be seen at Oxford in the collection of that naturalist.

16. Transparent, cubical zeolite, of a greenish tint. I found this specimen in the interior of Fingal's Cave, in the crevice between two prisms. It is therefore very evident that this little group of cubical crystals had been formed in that fissure in a very slow and insensible manner, by a juxta-position of the zeolitic particles held in solution in the aqueous fluid. The greenish

colour of this zeolite arises from the decomposition of the iron contained in the
basalt.

17. White semi-transparent zeolite in
octagonal crystals.

18. White semi-transparent zeolite in
crystals of thirty facettes.

Such are the most remarkable zeolites
which I have found in the isle of Staffa. It
is possible that the waves and currents which
daily wear away its coasts may eventually
bring other varieties to light.

19. Granite of a red ground, and of the
same texture with that of the ancient
Egyptian granite, but of a much less vivid
colour. This red granite is found in rounded
stones of a pretty large size among the rolled
lavas thrown by the sea upon that part of
the island where the currents have made a
considerable breach. As every thing in
Staffa is completely volcanic, it is evident
that these blocks of granite, which are not
very abundant, but of which the rounded
forms are the result of friction, have been
transported from a distance by currents; for
the neighbouring islands are equally volcanic.
The sea must certainly be terribly agitated

to raise these rounded granites to the height at which they are found on the isle of Staffa, among blocks of the basaltic lavas equally rounded, which the sea throws up during times of high tides and great storms.*

* [These granite boulders belong to the glacial deposits of the West Highlands. They may have been carried by ice from the red granitic tract of the Ross of Mull.]

CHAPTER V

Stay at Mr Maclean's. — Customs and Manners of the Inhabitants of the Isle of Mull

MR MACLEAN of Torloisk has erected a commodious habitation, in the modern style, without parade, but in which great neatness and quiet simplicity everywhere prevail. It commands a view of the sea, and of the isles of Ulva, Gometra, Staffa, Iona, and a crowd of rocks, which make navigation here so dangerous.

This house is situated on a dry platform without trees or verdure; so that to make himself a small kitchen-garden, Mr Maclean has been obliged to dig away part of the volcanic rock, on which he has put soil brought from elsewhere ; he shewed me several difficult and expensive operations of this kind which he had carried out. On my asking, why he suffered to remain standing upon the place a kind of large cottage built of dry stones, covered with thatch, or rather heath, and lighted by two narrow windows, which scarcely allowed the daylight to enter,

"It was there," Mr Maclean answered me with emotion, "that I was born. That is the ancient habitation of my fathers; and I feel inexpressible regard for this modest site, which reminds me of their virtues and frugal life." This reply paints the character of that estimable man better than all that I could say of him. It ought to be remarked, that Mr Maclean is a man of birth and fortune, that he has served in the army, made long voyages, and knows the ways of society. He has, notwithstanding, preferred his native soil, and an agricultural life, to a residence in London or Edinburgh, or the most fertile plains of England; so much does the dominion of our first habits cling to us when it recals the indelible remembrances of infancy.

Several ladies from Edinburgh, of agreeable conversation, were at Torloisk. One of them, a relation of the Melforts, of whom there is a branch settled in France, was a woman of talents and education. An officer, nephew to Mr Maclean, with two of his friends, was also on a visit at the house,

where all were united in the delightful bonds
of confidence and friendship.

Miss Maclean, an only daughter, pretty,
with a graceful figure, interesting from her
talents, her acquirements, and her modesty,
played extremely well upon the harpsichord,
and was in every respect the charm of
that society. She had attentively studied
the language, poetry, and music of the
Hebridians.

Miss Maclean assured me, in several
conversations which I had the pleasure of
having with her on the subject, that to one
acquainted with the language, the usages,
and the manners of the country, it was
difficult to conceive how the English writers,
who were strangers to the Celtic tongue,
should have so obstinately persisted in
doubting the existence of the ancient poems
of Ossian. She admitted, indeed, that they
are incomplete, and often altered; because,
handed from generation to generation, and
from bard to bard, they could hardly fail to
sustain some loss in passing through so
many hands. But she maintained that,
none the less, there exist entire pieces of
them, as well as the remains of the music

suited to them. It is a strange kind of music when compared to ours, but has a powerful charm for the Highlanders, since it reminds them of the combats, the victories, the loves, and the splendid deeds of their heroes.

No one could better convert the unbelievers than the lovely Miss Maclean, and I invite her in the name of the sister arts, of poetry and music, which she knows so well, to publish her researches respecting the poems and the airs of the ancient Caledonians. *

Mr Maclean's domestics, both men and women, were in the Hebridian dress. I have already described that of the men, in speaking of the inhabitants of Dalmally. That of the women is much less complex.

---

* Besides what Macpherson has said upon the subject, John Smith, minister of Kilbrandon in Argyleshire, has written in favour of the authenticity of the poems of Ossian, Ullin, Orran, &c. Mr. Nichol, of Lismore, has also treated of the same subject. John Clarke, of Edinburgh, has given a translation of the Caledonian bards. I purchased also at Edinburgh a collection of Gaelic music, engraved at the instance of a presbyterian minister, as well as many other printed and manuscript pieces relative to this question ; which I can communicate to such as it particularly interests. But this great question being foreign to the principal object of my researches, I shall forbear to expatiate longer upon it here.

Their long flowing hair, which is in general
black, forms the only ornament of their
heads. Some of them, probably from
coquettishness, keep it back with a simple
woollen fillet of different colours, among
which red and green are invariably predominant. Their chaussure is economical;
for they wear neither shoes nor stockings;
and in spite of the length of the winter, and
the incessant wetness of the climate, though
they go with their feet bare, and their heads
uncovered, they yet have very fine teeth.
Their dress consists of a bodice, or rather a
kind of vest, and a woollen petticoat in
large checkers of red, green and brown,
shaded with blue. This stuff is the general
and favourite material used by the
Highlanders, serving alike for men and
women. It is for the most part not
manufactured in the Highlands. It is
known as *Tartan.*

In England they eat very little bread;
but there were three different kinds used at
Mr Maclean's table. The first, which is a
luxury for the country, is sea-biscuit, which
vessels from Glasgow sometimes leave in
passing.

The second is made of oatmeal, kneaded without leaven and then spread with a rolling pin into round cakes about a foot in diameter and the twelfth part of an inch thick. These cakes are baked, or rather dried, on a thin plate of iron hung over the fire. This is the principal bread of such as are in easy circumstances.

The third kind, which is specially appropriated to tea and breakfast, in the richer families of the isles, consists of barley cakes, always without leaven, and prepared in the same manner as the preceding, but in leaves so thin, that after spreading them over with butter, one can easily fold them double; which is not unpalatable to those who are fond of this kind of dainties.

At ten in the morning, the bell announces breakfast. All repair to the parlour, where they find a fire of peat, mixed with pit-coal, and a table neatly served and covered with the following dishes:—

Slices of smoked beef.

Cheese of the country and English cheese, in trays of mahogany.

Fresh eggs.

Hash of salted herring.

Butter.

Milk and cream.

A sort of pap, of oatmeal and water [porridge]. In eating this thick pap, each spoonful is plunged alternately into cream, which is always alongside.

Milk mingled with the yolks of eggs, sugar and rum. This singular mixture is drunk cold and without having been cooked.

Currant jelly.

Conserve of myrtle, a wild fruit that grows on the heaths [blaeberry-jam].

Tea.

Coffee.

The three sorts of bread above-mentioned.

Jamaica Rum.

Such is the style in which Mr Maclean's breakfast-table was served every morning, while we were at his house. There was always the same abundance, and I noticed in general, no other difference than in the greater or less variety of the dishes.*

* Knox, who travelled more among the Highlanders of the mainland than among those of the islands, gives the following particulars of the breakfasts of the more wealthy families :—

" A dram of whisky, gin, rum, or brandy, plain or infused with berries that grow among the heath; French rolls; oat

At four o'clock, they sit down to dinner.
Here is the menu of the meal which I noted
exactly in my journal.

1. A large dish of Scottish soup, composed
of broth of beef, mutton, and sometimes
fowl, mixed with a little fine oatmeal flour,
onions, parsley, and plenty of pease. Instead
of slices of bread, as in France, slices of
mutton and the giblets of fowls float about
in this soup.

2. Black pudding made with bullock's
blood and barley flour, seasoned with plenty
of pepper and ginger.

3. Slices of beef, broiled; excellent.

4. Roasted mutton of the best quality.

5. Potatoes done in the juice.

6. Sometimes heath-cocks, wood-cocks or
water-fowl.

and barley bread; tea and coffee; honey in the comb, red and
black-currant jellies; marmalade, conserves and excellent
cream; fine-flavoured butter, fresh and salted; Cheshire and
Highland cheese, the last very indifferent; a plateful of very
fresh eggs; fresh and salted herrings broiled; ditto, haddocks
and whitings, the skin being taken off; cold round of venison,
beef and mutton hams. Besides these articles, which are
commonly placed on the table at once, there are generally cold
beef and moor-fowl to those who chose to call for them.
After breakfast, the men amuse themselves with the gun,
fishing, or sailing, till the evening; when they dine, which
meal serves with some families for supper."

7. Cucumbers and ginger pickled with vinegar.

8. Milk prepared in a variety of ways.

9. Cream with Madeira wine.

10. Pudding made of barley-meal, cream, and currants (raisins de Corinthe), and cooked in dripping.

All these various dishes appear on the table at the same time; the mistress of the house does the honours, and serves every body.

There is no delay in drinking the first toast; it is again the mistress who is charged with this ceremony. A large glass filled with port-wine is presented to her; she drinks the first to the health of all the company, and passes the glass to one of the persons who sit next to her; and thus from one to another the glass makes the round of the table.

The side-board is furnished with three large glasses, one for beer, another for wine, and the third for water, when any one asks for it unmixed, which is not often. These glasses are common to all at table; they are never rinsed, but merely wiped with a fine linen cloth.

The dessert, from the want of fruit, consists usually only of two sorts of cheese, that of Cheshire, and that of the district.

The cloth is removed after the dessert, and a table of well polished mahogany appears in all its lustre. It is soon covered with fine decanters of English glass, filled with port, sherry, or Madeira, and with large bowls of punch. Small glasses are then distributed in profusion to every one.

In England, the ladies leave table the moment the toasts begin. The custom is not precisely the same here; they remain at least half an hour after, and justly partake in this merry feast, wherein formality being laid aside, Scottish frankness and good humour have full room to show themselves. It is certain that the men gain thereby, and that the ladies do not lose.

We drank in particular to the health of each lady.

To that of the guests one by one, calling them by their names.

To the fatherland.

To liberty.

To the happiness of mankind in general.

To friendship.

We foreigners, drank more than once to our good friends the Highlanders; and the company answered in full chorus with drinking "To your friends in France," and in a lower tone, with a glass of mild Madeira, "To your lady-friends."

The ladies then left us to give their orders for tea. They were absent but a short time and returned in about half an hour after. The servants then brought in coffee, slices of buttered bread, butter, milk, and tea. Music, conversation, reading the news, though somewhat old by the time they reach this, and walking, when the weather permits, fill up the rest of the day which is thus quickly brought to a close. But, what is perhaps a little unpleasant, is that at ten o'clock one must again take one's seat at table and share until mid-night in a supper of nearly the same kind as the dinner, and in no less abundance.

Such is the life that is led in a country, where there is not a road nor a tree, where the mountains are covered only with heath, where it rains for eight months of the year, and where the sea, always in motion, seems to be in perpetual convulsions.

The winter is cold only for about two months in the Hebrides, and the snow lies there only a brief time, but winds and rains prevail during the greater part of the year. Neither wheat nor rye can be grown; barley and oats, however, thrive here, and are reaped in the month of October, though it is necessary to dry the grain in kilns, without which precaution, it would shoot and could not be ground.

The greatest part of the barley is fermented and distilled, to procure a spirituous liquor in which the people delight. It is called whisky. The meal of the oats serves for making cakes.

The isle of Mull is hardly more than from twenty to twenty-two English miles long, and fifteen or sixteen broad; a mile is termed in the Hebrides *scocs* [?]. There is no trace of a regularly built village in the whole island; the houses being almost always scattered apart, both on the coast and in the interior. They are built of irregular blocks of basalt ranged, without much attention to order, in walls of great thickness; for materials of this kind are plentiful, and always within reach of the builders. The height of the walls scarcely exceeds five feet, and that of the entrance is usually only three feet. The islanders in more easy circumstances adapt a door to it; but the greater part of the inhabitants do without any. The roof is often covered with flat stones, upon which are placed clods of turf. But those who have the means to procure some wood make use of it with a thatch of heather or oaten straw, fastened and kept down by long ropes of heather to protect it from the violence of the winds.

The fire-place is always in the middle of the hut, and the smoke of the peat escapes by a hole in the roof, which is a little to one side, that the rain may not extinguish the fire. The Esquimaux and Luplanders display much more art and industry in lodging themselves.

The islanders of Mull go bare-footed and bare-headed, fearing neither rain nor hoar frost. Fathers of families sometimes have a Scottish bonnet, and the married woman a head-dress of coarse cloth. But all the young folk, both girls and boys, go about with their heads bare, and without shoes or stockings. I am speaking of the common class of people.

Almost all are fishermen or shepherds and each cultivates some small spot on which barley or oats, and some potatoes are grown; potatoes with milk form their principal nourishment. Those on the sea coast, or within reach of lakes, have the resource of fishing. They catch salmon, which they dry in the smoke, and herrings, which they sell, and from which they get the oil for their lamps.

The more intelligent enter the English navy, and make robust, sober sailors, familiar with all the dangers of the sea.

The population of the island is about seven thousand.

There are three parishes, nine churches, five schools; the religion of the district is the Presbyterian.

The women here, are in general, small, ugly, and ill-made—the natural consequence of toil, bad food, the want of good clothing, and the inclemency of the climate. I saw two or three who were not so bad-looking, and whose figure was even rather comely, but these belonged to families in a more comfortable condition. The sun being almost always hidden under clouds, or enveloped in mists, the complexion of the women would be very pale, were it not discoloured by the peat-smoke, amidst which they pass their lives in huts without chimneys.

Notwithstanding the wetness of the climate, I could never perceive that the custom of going bare-headed was injurious to the teeth. Both men and women have very fine sets, and are in general, especially the men, very healthy. The disorders which might be expected to arise from the frequent rains of this climate, are mitigated by a temperate life, hard work, and the purity of the air. The whole of their sustenance consists only of a milk-diet, a few potatoes, fish at certain times of the year, and oatmeal made into porridge or cakes. Their drink is pure water; and some drops of whisky on their festive days give them their greatest happiness.

I made some enquiries in Mull respecting the age of the old men. Mr Maclean of Torloisk assured me, that a man of his acquaintance who resided in the neighbour-hood of Aros, had died about seven years before at the advanced age of one hundred and sixteen years, and that there were a number living above eighty; but these belonged to the class who were in comfortable circumstances.

There are no horses in the island save those of a small race; the black cattle are also very small, but delicate when fattened; they are exported to England, and form one of the principal revenues of the isle of Mull. There are also in it two kinds of sheep which I shall soon describe, some goats; no pigs at all,* and few fowls owing to the difficulty of rearing them. At Aros, at a house by the shore, I saw some geese, some domestic ducks, and three turkeys; but the heads of these last were pale, and I doubt whether they can thrive there.

The high mountains produce deer, though few in number, and of a smaller size than the common kind. Heath-cock, of the greater and lesser species, are plentiful; there are also some wood-cock, but no hares. The only small bird which I saw in my journey, was the ortolan.†

* [There was a traditional prejudice against pigs and pork which is even now by no means extinct in the more sequestered parts of the Hebrides, where the influence of Saxendom has been least aggressive.]

† [The author was fortunate if the bird he noticed was this bunting, of which not many specimens have been seen in Scotland.]

The island is now denuded of trees, though formerly it must have had many of them; as may easily be seen among the peat-mosses; for when the best of these are dug to a certain depth, it rarely happens that roots and stumps of beech, pine, and birch are not met with. I even believe that if the trouble were taken to plant evergreens and birches, they would succeed. I base this opinion on having seen a small thicket near Auchnacraig, at the end of the island opposite to that of Torloisk.

The level country and the mountains are in general covered with heather and grass.

The tides rise to a great height in these parts, and the shores abound with sea-wrack, which has for some time back been burnt for its alkali, which merchants from Glasgow come to purchase. But this useful object of industry is exclusively engrossed by the lairds, or a few wealthy proprietors. The sea-wrack, when fresh, is used with success as a manure.

There are yearly exported from Mull about fifteen hundred head of black cattle; but, from their small size, they bring only about three pounds sterling each.

*Of the sheep of Scotland, particularly those
of the Isle of Mull*

As I was in a position to obtain in this
spot some exact information on this subject, I
shall here state it as briefly as possible, with
the intention of being useful to those who
occupy themselves with this great subject of
national economy.

In the mountains of Scotland, and the
Hebrides, there are only two kinds of
sheep; the original race of the country,
which are small, but of very good quality,
and a breed introduced from England, much
larger and known as *English sheep*.

The wool of the Scottish sheep is much
superior to that of the English sheep, and
even approaches the Spanish wool in fineness. But many people prefer the English
sheep because they yield a fleece double that
of the Scottish, are fatter and fuller of
flesh. They accordingly sell at a much
higher price.

An English sheep, in good condition, sells
upon the spot for half a guinea and often
twelve shillings; whilst a Scottish one seldom
fetches above six or seven shillings.

The wool is sold here by weight; twenty-four pounds being called a stone; this quantity is generally valued at from six to seven shillings. The pound contains sixteen ounces.

During summer and winter the flocks of sheep range the mountains or valleys, both night and day, without any shelter; yet the extreme wetness of the climate does not in the least injure them.

They never have any fodder during the winter, not even when there is snow; but it should be observed that in these isles, though so far north, the snow does not lie long. By a very rare occurrence, however, in the winter of 1783, it remained on the ground for two months; during which time the sheep browsed upon the tops of a tall kind of heather, which is plentiful in the country and which then stood out above the snow. The poor animals, however, suffered severely during that long winter, and became very thin. But a much greater number of them died from accident, than from want or disease; and on the reappearance of the grass they recruited very fast, and fattened as usual.

The rams are carefully separated from the ewes in the month of September, and are not admitted into the flocks again till the twentieth of November, that the lambs may be yeaned only in the best season.

The ewe brings forth and takes care of her lamb without any assistance. The shepherd, who, from time to time, visits his numerous flock, to prevent the animals from straying too far, or endangering themselves on slopes too steep for them, sees the young lambs, who soon run after their dams and crop the new-sprung grass.

In the third month the lambs are separated from their mothers, being then strong enough to do without them, and they are formed into flocks which are put into separate enclosures, under the care of a shepherd.

When a continuous pasture-ground is of great extent, one man and two dogs are sufficient to keep fifteen hundred sheep. But when the area of pasture is considerably less, and it is desired that the animals should feed more regularly, a shepherd and two dogs are requisite for every eight hundred.

Nineteen or twenty rams are sufficient for eight hundred ewes. Each shepherd comes every evening to sleep in a cabin built of stone, nearly in the centre of the pasture-ground.

The sheep are exempt from all disorders except two, the pleurisy, which is rare, and the staggers, which makes them reel round, and always ends in their death. This disease is unfortunately pretty common, and often attacks those animals which appear to be the most healthy.*

Salt is never given to sheep in the Hebrides; although the people are aware

* This disease is the same as that known in Tuscany by the name of *Pazzia*. The animals affected with it, and which are called in France *moutons lourds, moutons imbecilles,* waddle in their walk. Abbé Fontana, in a letter upon this subject addressed to M. Darcet, and inserted in the *Journal de Physique*, tom. i. page 227, 1784, says, " it is very remarkable that the sheep attacked by it, generally fall on one side, and that the vesicle which occasions it is found to be in the lobe of the brain opposite to the side on which they fall. This observation holds good in all cases; and the animals constantly fall on the same side." The celebrated physician of Tuscany, from several microscopic experiments on the liquor contained in these vesicles or *hydatides* in the brain of sheep, concludes, that " the particles, which are seen floating in that liquor, are *real animals.*" This new and singular discovery, says that philosopher, "may throw light on some disorders of the human brain, and even on insanity; since vesicles as large as a pea, and sometimes larger, have been found in the brains of persons who have died of that malady, so terrible and humiliating to man," page 231 of the same paper.

that it would be useful to them ; but their flocks being very numerous, and the salt, on account of government duties and carriage, pretty dear, the expense would be too great. Were it not for this hindrance, the inhabitants would undoubtedly use it for the fleecy race ; for they are well aware that cows and oxen, that are pastured upon herbage washed by the sea, thrive well, become fat, and have sleek coats.

Here I ought to mention a practice observed in several parts of the north of Scotland, particularly in the lower pastures, though not in the Hebrides or the higher parts of the Highlands, that of tarring the sheep. I think this fact will naturally find its place here.

The owners of the numerous flocks of these districts, where the winter is more severe than in the Hebrides, believe that the intense cold gives the mange to the sheep ; and as a security against this, they make use of the following preservative :

In the month of November each shepherd takes, for instance, two barrels of tar, one barrel of butter ; or a larger quantity of each, according to the number of the flock,

but always in the proportion of two-thirds
of tar to one-third of butter. These two
substances are then melted or boiled together
in a cauldron; when the mixture is completed and cooled, each sheep, tied up by the
feet and stretched on a hurdle, is rubbed
over with the composition. This operation
is performed by separating the wool into
thin flakes so as to lay bare the skin and
thus avoid as much as possible soiling the
fleece.

In the opinion of the sheep-farmers of
the country, this practice has two advantages.
The first is, to keep the animals in a state
of health; the second, that of making them
yield a greater quantity of wool. The
wealthiest sheep-farmers, whom I had an
opportunity of consulting upon the subject,
assured me, that the wool was certainly more
abundant when the sheep were tarred. It
is true, they said, that this wool sells for
one-half less than unsmeared wool, because
being full of earth and other dirt, it is heavier.
The operation of removing the tar, consists
in soaking the fleece after it is shorn in
warm water, into which butter has been
melted. But this process must be expensive;

and the wool never reaches a quality equal to that of the pure wools.

Five thousand sheep require twenty barrels of tar, and one-third that quantity of butter. This expense appears at first much more considerable than it really is; for, on dividing it among five thousand, it does not amount to five *sols* [2½d] a-head. Besides, this anointing with butter and tar gives the sheep a sort of artificial grease, which supplies what the rigour of the climate deprives them of; and if it tends to keep these useful animals in better health, and also to increase the quantity of their wool, however ridiculous, apparently expensive and difficult of execution by those unaccustomed to it, it may at first appear, the practice is yet, perhaps, worthy of some attention and careful examination on the part of those who are more particularly engaged in so important a branch of economy.

CHAPTER VI

*Departure from Torloisk.—Stay at Aros.—
Visit to two worthy Farmers, the
Brothers Stuart of Aros.—Excursion to
the Mountain of Ben More, the highest in
the Isle of Mull.—Stop at Mr Campbell's,
of Knock.—His agricultural Operations.—
The curious Lavas which his clearings
have brought to light.—Departure from
Aros for Auchnacraig*

I WAS treated with such engaging marks
of politeness and kindness by Mr
Maclean and all his family, as well as by
his visitors, that it was impossible for me to
leave them without regret and gratitude. I
should be glad to prove to them that I shall
never forget them. This estimable philosopher kindly accompanied us for several
miles on our return.

During my stay at Mr Maclean's I took
a survey of the volcanic hills near his house,
and pushed my walks to right and left,

along the coast where the sea has laid bare great cliffs, well fitted to reveal the structure of those currents of material formerly a prey to the subterranean fires. I shall give a description of them in the chapter which will be devoted to the mineralogy of this island.

We set off, mounted upon little half-wild horses, and on the same day reached Aros. Here we remained next day in a miserable lodging, where we found only some barley meal, which was made into pottage for us with milk, a little smoked salmon, some sheep-trotters; no wine nor beer; but whisky, which scalded our throats, and, to crown all, bad beds. Our host, however, was a worthy man, and made incredible exertions for our accommodation. That satisfied us, and he even held out the hope of some fresh fish for next morning.

My companions were left to enjoy the fish, for, as for myself, I had determined to set out at sunrise to visit the high mountain of Ben More, and William Thornton, whose taste for Natural History was growing, wished to accompany me.

It is counted scarcely three miles from Aros to Knock, by a pretty good road,

surrounded with picturesque landscapes, but
of a somewhat wild type.

We saw on a broad tract of pasture in
the bottom of a small valley, washed by the
sea, one of those columns called *Cairns*, of
which I could not take the dimensions, for
the ground was under water at the time.
But as near as I could judge by the eye, it
might be about fourteen or fifteen feet in
height, and seemed to consist of sandstone.
One is always astonished to see how common these ancient monuments are in the
Hebrides and on the mainland of Scotland.
Popular tradition traces them all back to the
time of Ossian, which is merely to say, that
their origin is lost in the night of bygone
ages.

The house of Mr Campbell, of Knock, is
agreeably situated at the foot of a high
mountain, and not far from an arm of the
sea, very plentiful in fish. Mr Campbell
was then at Oban, but we were received
with affability by the mistress of the house,
who offered us tea and rum. We begged
that she would procure us a guide to direct
our way to the top of Ben More; but her
son, a youth of seventeen or eighteen years

old, was willing to accompany us himself. This young man, who had a very good figure, and was dressed in the Hebridian costume, immediately presented us with fowling-pieces, saying, that he had excellent dogs, and that we should certainly find some *black-cocks*; for he could not imagine that we should wish to climb so rugged a mountain, for any other purpose than the pleasures of the chase, which he passionately loved himself. He was, therefore, much surprised when I took out my hammers, and told him, that I had come to examine the stones of the place. He then shewed us immense heaps of them which had come from a considerable clearing that had lately been made by his father in the midst of the lavas. All these stones, broken into fragments, were afterwards used to form fences round a piece of ground which required much labour, time and expense, to reclaim. A larger collection of lavas is seldom to be met with. I shall presently allude to it.

As we intended to return to Aros in the evening, we lost no time in beginning to scale the steep sides of Ben More. In my journeys among the High Alps I never found

so much difficulty as here. Almost impenetrable heather, above a soil saturated with
water, covers the lower ground, the middle
and the summit of the mountain, which rises
in the shape of a sugar loaf. It is impossible
to make any progress save by following the
small gullies which the waters have worn,
and these narrow and steep tracks are, as it
were, so many threads of water in the midst
of which one has to walk. A black and
bushy heath covers, as with a funeral pall,
the stones which might interest the naturalist
and lighten his fatigues. Not a plant, not
a moss is to be seen, every thing is here
smothered by this voracious heather.*

The stones which some of the deeper
gullies have laid bare, and those which have
been broken off by frost, are all volcanic.
But they present no variety; all of them
are whitish-grey lavas, mingled with some
spots of zeolite.†

* [Long heather is undoubtedly sometimes an impediment
in mountain-climbing in Scotland. But had the author been
accustomed to it, and been able to attend to other aspects of
the vegetation of Ben More, he would have found some
interesting plants on that mountain.]

† [The struggle against the heather seems not only to
have hidden the flora from the eyes of the traveller, but to
have concealed also the interest of the rocks.]

I had reached a great height, when, wearied with seeing only the same lavas, and meeting with no other plants than the toilsome heather, whence started from time to time some black-cocks, which young Campbell shot with great dexterity, I resolved to go no farther. But William Thornton braving every thing and desirous of gaining the highest summit, climbed to it. The stones which he brought down with him afforded no variety. Upon the whole, the mountain of Ben More, notwithstanding its height, and a kind of resemblance which it has at a distance to Mount Vesuvius, does not repay the trouble of ascending it.* We gladly returned therefore to rest ourselves at Knock, where the lavas are much more interesting, and where I made a collection of specimens of them. We then took leave of young Campbell and his mother, notwithstanding their pressing invitation to stop. We were expected at Aros, and thither we returned.

* [Apart from its exceedingly interesting geology, and the commanding view from its summit of the structure and disposition of the central volcanic high grounds of Mull, Ben More (3169 feet) is well worth ascending, for the magnificent panorama which its top affords of the greater part of the western Highlands and Islands of Scotland.]

It was determined that we should set off for Auchnacraig at ten next morning. This was a distance of eighteen or twenty miles, which we were glad to accomplish by land, as we should thus have an opportunity of examining that part of the island, and at the same time avoid the stormy Sound of Mull; for from Auchnacraig we could next morning easily reach Oban to breakfast.

We left Aros at the appointed hour; but first had the pleasure of breakfasting, by invitation, with the Messrs Stuart of Aros.

These gentlemen are two brothers, who have a commodious habitation on a small hill, where they have successfully cultivated pasturage, barley, oats, and potatoes. In that modest asylum, free from care and disquiet, they pass away their days with a happiness which ambition has never tasted. Two intelligent and industrious sisters share with them in the cares of the household. Here they enjoy all the pleasantness of rural life; I would only wish for them, and I do it with all my heart, a more favourable sky, and a soil better suited to their tastes and to their attainments in agriculture.

We took leave of them at ten in the
morning of the 29th of September.

A few miles from Aros, near the edge
of the sea, we observed the ruins of a
catholic chapel, where are still visible a
gothic bas-relief in sandstone, representing the Virgin Mary between two little
seraphim, and a large grave-stone on which
may be seen the effigy of a warrior in
complete armour, that is, with helmet,
armlets, cuisses, buckler, and sword. One
of our guides told us, that it was the figure
of a hero of the clan Maclean. Beside
this sepulchral relic, we observed another
representing, also in relief, a woman of tall
stature, dresssed in the gothic style of the
ancient ladies of France. The name of the
place where these ruins are to be seen is
Galchayle.

Thence we continued our course, rather by
paths than by roads, to Lenigorn,* Ardmitrail,
and Corinahinish.† It must not be supposed
that all these names indicate villages, or even
hamlets. Alas! they are applied only to
some huts, scattered at distant intervals
amidst these dismal deserts.

* [Pennygown?]   † [Corrynachenchy?]

Every thing along this road is volcanic; but the compact homogeneous grey-coloured lavas do not present much interest. They are besides so thickly covered with mosses or lichens, that it is necessary to break them before they can be distinguished.

It was only in the neighbourhood of Ledkirk that I found some hard compact lava, disposed in slabs, which pleased me. It is a kind of white lava; and at first sight might be taken for a fine limestone of that colour, but which, on more attentive examination, presents vitreous characters that leave no doubt that it is only a basaltic lava bleached in the neighbourhood of some solfatara, or by remaining long in a liquid impregnated with some acid. It is remarkable that these lavas have preserved their magnetic property. I made a collection of them in order to compare them with similar specimens which I found on Mount Mezinc, in Vivarais, and with those of the extinct volcanos in the environs of Padua, and of the Euganean Mountains.

From Ledkirk we passed on to Garmony, and thence to Scallastle, leaving the little fort of Duart on our left. On a green rising

ground near Scallastle, we saw a large
druidical circle, formed of huge blocks of
rough granite. After halting at this ancient
altar or temple, we quitted it promptly and
with indignation, as we reflected that these
cruel priests, of a still more cruel religion,
had perhaps sacrificed here some Iphigenia,
thrown by a tempest upon this new Taurica.

We arrived in the evening at Auchnacraig. This is the name of a small creek,
where there is a solitary house, very
wretched, very smoky, having, however, two
stories, and with chimneys. One could not
tell at first whether it was a farm-house or
an inn; it turned out to be both. The
breadth of the arm of the sea which
separates this part of the Isle of Mull from
the mainland of Scotland at Oban, being
inconsiderable, the ferry is much frequented
for the transportation of cattle; and this
house affords shelter to those who are driven
in by bad weather, or whom commercial
errands call to the island. One lives here
after the manner of the Hebridians, that is
to say, very frugally; but, for the rest, the
landlord was a good sort of man, very
inquisitive after news, a lover of antiquities,

and having as much veneration for Fingal
and for Ossian, as the Jews have for Moses.

Next day, heavy rain having come on,
we did not venture out of doors; I took
advantage of the time to label my specimens,
and put my journal in order.

On the succeeding day, the weather was
less rainy, but the sea was very boisterous.
We made some rambles in the vicinity; and
half a mile from the hostelry we noticed a
bank of limestone adjoining a bed of sandstone, and both of them inclosed in a
current of lava.

At a short distance from this bank, I
observed a large unhewn column of sandstone, lying flat on the ground, and broken
in the middle. On measuring, I found it to
be twenty-one feet long. Our host, who
accompanied us, did not fail to make us
admire this ancient monument. "Never
was there a person, except Ossian," said he
quite seriously, "who could move so enormous a stone. The weather, or some
earthquake has overthrown it, and nobody
in the island can now set it up again."

On the next day, 2nd October, it rained
all the morning; but towards the evening it

cleared up for a little. Count Andreani, who wearied in so dismal a solitude, and so bad a lodging, resolved to take advantage of this interval of calm to cross over to Oban, where he should wait our arrival. The only craft in the place was a small skiff, in bad condition, with two rowers, who might be looked on as children, for the eldest was not more than fourteen years of age. The wind was variable, and the sea not very smooth. In vain, however, did I represent to him that it would be better to wait till the next morning; nothing could prevail on him to stop. He set off in the skiff, with his two servants, at half-past four, telling us, that he should sleep in a good bed, and eat a better supper than we, at the house of the brothers Stevenson, of Oban, where he expected to arrive by seven the same evening.

Less adventurous, than he, and perhaps more prudent, I persuaded William Thornton to remain with me at Auchnacraig till the sea became more moderate. So we wished our friend a good passage, and followed him with our eyes as long as we could see him. We then turned slowly back to our wretched and dreary habitation. I wrote till eight; we then supped, and I went to bed at ten.

A strong gale then made itself heard, accompanied with a great deal of rain; but I was not anxious about the situation of our companion, who must already have reached Oban.

I had scarcely fallen asleep, however, when a loud noise awaked me. I heard a rapping and calling at the door; I rose, and after rousing the people of the house, who went and opened it, we saw our poor friend Andreani enter with his attendants, miserable and as completely soaked as if they had been repeatedly plunged under water. They were caught by the storm, when half way over, the tempest had several times driven them near to Oban, without their being able to make the harbour. The night was so dark, that it was impossible to know where they were, and it was not until after they had run the greatest dangers, and so to say by mere chance, that they regained the little haven of Auchnacraig.

They were numbed with cold; our first care, therefore, was to warm them. A

large fire was lighted, rum and tea were
given them to drink, and everything went
well. Count Andreani was the first to
laugh at his adventure; but it was no
laughing matter to his two servants, who
having never before seen anything but the
fertile and smiling fields of beautiful Italy,
felt themselves rather out of their element
here. They had been so deeply impressed
with the danger and the frightful spectacle
of a stormy sea, amidst the darkness of night,
that after having a thousand and a thousand
times, returned thanks to the Blessed Virgin,
whom they had invoked, and who had saved
them, they raised their hands to heaven, and
swore never again to leave this island,
all barren as it was. " We would rather,"
said they, " browse the grass, than expose
ourselves again, to the fury of that abominable sea." They then grumbled at their
master for his imprudence and madness, in
coming to visit the most detestable country
in the world. Their gestures, their panto-
mime, the play of their features, and the
serious tone of their lamentations, afforded
me a truly comic scene.

Repose for the rest of the night, and fine

weather next day, partly effaced the impressions of the preceding evening. The
sea, however, was not yet navigable, so
what was to be done?—to be busy; that
is the surest way of chasing away ennui.

At sun-rise, therefore, I took one of those
walks from which one always gets some
benefit, whether in instruction or in health,
and in which I always find my advantage
wherever I travel.

As we were coming into Auchnacraig
on the day of our arrival, a great black
rock, vertical and nearly isolated, had riveted
my attention. I suspected that it might be
a basaltic colonnade, and I wished to make
sure of this conjecture, and after a walk of
a mile and a quarter, I arrived at the foot
of one of the most astonishing productions
of the volcanic fires that I ever had an
opportunity of observing.

It was a kind of circus, of the antique
kind, formed of natural walls of basalt rising
vertically with such a regularity, that at a
first view, the spectator can hardly persuade
himself that it is not a work of human industry and art. But the utmost stretch of
human force, helped by mechanical means,

could never have been capable of placing such enormous masses in their position. The whole must be regarded as the effect of a vast conflagration [incendie], which, instead of destroying, has here produced results analogous to those of a creative power.

This grand natural monument excited a just admiration and even enthusiasm in my mind. I could not tire gazing at it, and I remained more than two hours in beholding, studying, and observing it over again from different points of view. I went in quest of my companions, who were delighted at the discovery and admired no less than myself those vast basaltic walls, standing alone, and rising boldly and vertically around a circular space, which formed an arena, well fitted for races or games in the ancient style.

It is no less remarkable, that the accessories of this singular production of the subterranean fires seem to have been designedly placed in the vicinity, to furnish the key to the problem of its formation.

I took, with the most scrupulous care, all the measurements of the height, the thickness of the walls, as well as the size of the circular enclosure.

On the 4th, I visited it once more.
In the afternoon of that day, the weather
beginning to assume a more settled appearance, Count Andreani said to me, that he
was resolved to try his fortune once more,
and that he should set off at four o'clock;
which he did. This time the wind was
favourable, and as the little boat could not
carry us all, we suffered him to start,
promising that we should soon rejoin him.

He sent us back the boat, during the
night, with a supply of eatables, for our
cheer had been very meagre for some days,
having exhausted almost the whole stock of
provisions at Auchnacraig.

This supply was extremely useful, for the
weather again became bad next morning,
and the sea ran too high for us to trust ourselves upon it in so frail a skiff. I employed the time in new excursions, and in
arranging my notes, particularly those which
related to the mineralogical history of the
isle of Mull. These I have thrown into a
separate section, that such of my readers as
are interested in that science may find the
objects which refer to it, united in one
chapter, and that those to whom the subject

may be indifferent or tiresome, may easily skip that part. It may not be improper to mention a second time, that this will be my ordinary mode of procedure.

At length, on the evening of the 6th of October, a bark having come in to Auchnacraig with a cargo of beeves, and having to return on the morrow, we resolved to take the opportunity thus offered; we accordingly embarked at six next morning, not for Oban, but for the isle of Kerrera, where we landed at eight. We crossed the isle, which is not very large, and at its end found a boat, which took us in less than an hour to Oban, where our dear Count Andreani awaited us with the keenest impatience.

CHAPTER VII

Natural History of the Island of Mull

THIS island, though one of the largest
of the Hebrides, is not more, as I
have already said, than from twenty to
twenty-two miles in length, by fifteen or
sixteen in breadth; but being of a very
irregular shape, it may be computed to be
at least eighty miles in circumference.

I shall proceed to describe the parts
which I visited, in the order of my journey.
Those who would wish to explore the island
in the same pursuit, by disembarking at
Auchnacraig, and coming back to Aros, by
land, along the left side of the Sound of
Mull, will trace my itinerary by beginning
where I have ended.

AROS

ROAD FROM AROS TO TORLOISK

*Columns of Basalt. Lavas, compact, black,
grey, reddish, intermixed with kernels of
white zeolite. Blocks of rolled Granite
on the top of some basaltic mountains*

THE ancient castle of Aros, once the residence of the famous Macdonald of the Isles,
is now nothing more than a ruin. Its
remains stand on a small colonnade of basalt
by the margin of the sea, and on the right side
of the entrance into the small bay of Aros.

The Aros river, which might with more
propriety be called a large brook, takes its
rise from a marshy tract, about the middle
of the island. From its source to its mouth
it runs on a compact lava, which varies, in
colour from a deep black to grey and
reddish. This lava is in general hard and
compact; some streams of it, however, are
friable and gravelly.

These compact lavas contain, in general,
so great a quantity of nests of white zeolite,
that this last material may be said to form
nearly one-third of the weight of the lava.

The zeolite occurs here in globular form, in general about the size of a pea. Some of these kernels are radiated; more frequently they are crystallized in a confused manner, and without any determinate form. I found nothing of this kind interesting for the cabinet, from Aros to Torloisk, and the reason is simple, for everything there is so covered with moss, lichens and heather that one is forced to confine one's researches to the bed of the small river, and to some ravines connected with it, where the rock is a little exposed.

In proportion as we approach Torloisk, at the distance of about three miles from the house, we find mountains entirely volcanic, and at least two hundred and fifty toises high. One is much astonished, in passing along their summit, to observe large blocks of granite, rolled and partly rounded, isolated, and resting on the volcanic material, to which, however, they do not in any way adhere, having evidently been transported hither by the effect of some great revolution. For adventitious bodies of this kind, and of so great a bulk, found on mountains and in an island where there is no solid mass of granite, incontestably proclaim that they have been deposited there by some great force.

Volcanic explosions, for instance, at the time when great conflagrations devastated these regions and formed these groups of islands which appear to have had the same origin, might have thrown out these blocks, torn off from granitic masses which perhaps exist at great depths beneath these ancient volcanoes.

It is within the verge of possibility that the higher parts of the mountains where the blocks of granite are now found, did not then form an elevated summit, but were rather part of the sea-bottom, where currents rolled these granitic blocks from a distance; it is possible, I say, that in these circumstances subterranean explosions, as terrible as those which could upraise the isle of Santorin, in the Archipelago, or Monte Nuovo, in Italy, may have equally uplifted the bottom of the sea and turned it into a volcanic peak.

Or, if the idea should appear more plausible, we may have recourse to the periods of those grand and ancient revolutions, when

mountains still higher were entirely sub-
merged ; a fact, which cannot be doubted,
since marine bodies are found in great abundance in beds of limestone or in argillaceous
strata, on the Alps or Apennines, at heights
three or four times greater. But all this
would deserve to be worked out in such a
manner as the nature of this book does not
permit me to attempt here.*

TORLOISK

*Black Basalt with and without zeolite ;
altered Basalt, which has lost its hardness and colour ; Basalt overcalcined,
of a blood-red colour, having the appearance of an argillaceous bole*

AT a little distance from Mr Maclean's
house, near the road leading to the sea on
the side of Kilninian, the shore is fringed
with scarped vertical rocks which are battered
by the waves and weakened by the frequent
rains, so that being quite bare, they can be

* [As pointed out in Chapter IX. of the first volume (p. 221)
the problem presented by the dispersed erratic blocks of the
Highlands cannot be solved by reference to volcanic agency,
but becomes intelligible when the transporting power of the ice
of the Glacial Period is recognised.]

studied with the greatest facility along the whole of that coast.

This great escarpment which extends as far as Loch-mari, is composed of different streams of basaltic lavas, of a deep black colour. Some of these streams are in irregular masses, others in tables, while some have assumed the prismatic form. Here I found some pretty large specimens of fine zeolite, of which some were crystallized in cubes, others in divergent rays, while a few were a little chalcedonic. They are usually very white, but some are fawncoloured from the decomposition of the iron, and others may be found of a light greenish tint. They are most frequently seen in large lumps enclosed in the lava; but those which have taken the cubical form, are oftenest found in the fissures which separate the different streams of lava.

The traveller should not omit visiting on the opposite quarter, that is, towards the path on the left in leaving Mr M'Lean's house to go to the shore, a quarry, out of which all the stones of his buildings have been taken. Here several lavas may be found which are worthy of attention, and can be examined with the same ease as the preceding, seeing that interior of the volcanic eminence has been laid bare by the quarrying operations.

The upper beds of this quarry consist of a black hard compact lava, containing kernels of white zeolite.

Those immediately below, probably acted upon by sulphurous acid, have lost a part of their colour and their hardness. They are grey, whitish, and most frequently of the colour of iron-rust. The zeolite which is there found enclosed, while preserving its forms and chemical properties, has notwithstanding assumed various tints.

Other beds, still lower, have undergone a different kind and much more considerable degree of alteration; they are bright-red, and contain, as well as the lavas above them, kernels of zeolite, unaltered in composition, but only softer and a little coloured. The lava itself has lost its hardness.

The lavas of this quarry, though they possess an identical base (pâte) and composition, have undergone different modifications, as much from the emanations which rise from this burnt ground, as from the action and effects of a long continued fire.

The different solfaterre * afford a constant and remarkable example of the active operation of vapours, not only on the colours, but also on the grain and the hardness of lavas, the constituent parts of which they disunite, and form out of them new gypseous, ferruginous, aluminous, sulphureous and other combinations. I have proofs, likewise, that the mere action of a long sustained fire, can, in certain circumstances, transform the hardest and blackest lavas, basalt for example, into a state of red calx, if I may use that expression.†

These supercalcined lavas in losing their first colour, lose also that which constitutes their hardness; and there are circumstances in which they become soft to the touch, and soapy like greasy clays. I have described a variety of this kind in my *Minéralogie des Volcans*, page 395, No. 10.

It is, therefore, of essential importance to distinguish accurately the two kinds of alteration which I have mentioned; the one is

* [Volcanic orifices from which steam, gases and other hot vapours ascend.]

† [Calx was an old chemical term to denote the powder or friable substance that remains after a mineral or metal is roasted or calcined, and its volatile parts are driven off.]

owing to the action of acids, and the other
to that of a long sustained fire. Thus the
black lavas which form the first beds in
the volcanic quarry of Torloisk are nowise
altered. Those which succeed them, and
which are grey and whitish, seem to have
been bleached and altered by acid vapours;
whilst the deepest strata, those where the
lava is of a blood-red colour and friable texture, appear to have suffered that modification
solely from the long continued action of fire,
and a veritable supercalcination. The fire
has not been violent enough here to change
this lava into a vitreous substance; but its
prolonged operation has disunited the com-
ponent parts of the rock and has rusted and
oxidised its ferruginous particles, which have
been changed to a red colour, like that of
the calx of lead, which a very violent and
long applied heat, converts into minium of
the finest red colour. The zeolite in kernels,
found both in the upper and the lower beds
of the Torloisk quarry, that is, both in the
black, and in the grey or whitish lavas,
is the same. It is the same also in the
deepest beds, where the lava is more altered,
and has become red. Even there this zeolite

differs from that of the other beds only in being a little softer; but the difference is not very perceptible.

What has taken place here with the zeolite, has happened also with respect to the acicular schorl in a lava of *Chenavari* in Vivarais, where the black schorl remains almost untouched in the midst of a lava that has been altered and has turned to a red colour by the continued action of a fire, which though strong has not been powerful enough to make it pass into the condition of glass.*

* The following is [a translation of] the passage in the *Minéralogie des Volcans* :—"Argillaceous basalt of a blood red, with specks of black schorl in the most excellent preservation, though the lava itself is changed completely into an argillaceous material, soft and soapy." *Op. cit.* p. 395, No. 10, Paris, 1784.

I ought to add here, that in saying that the lava had changed into an argillaceous material, I did not suppose that it had passed into the state of real clay. I merely meant that the lava thus altered has put on the external character of clays ; that is, that it is tender, earthy, and soft to the touch. I am glad to explain myself upon this subject, because several naturalists, who have written upon volcanos, have taken these substances for real clays, regarding them not merely as lavas reduced to an earthy condition, but as clays in place which have been burnt by the subterranean fires. In these cases, however, the schorls, chrysolites, zeolites, and even kernels of porous lava which are there enclosed, remove every doubt of the identity of these altered lavas with those in their neighbourhood which most frequently overlie them, or alternate with them, and which are perfectly sound.

I have quoted in the same work, which I published in 1784, on the *Mineralogy of Volcanos*, an example of this supercalcination daily effected even artifically. In Vivarais, and also on the other bank of the Rhone, near Montelimar, limekilns are constructed with their interior lining of very black and hard basaltic lava. The pit-coal fuel, with which these kilns are continually fed, has quickly vitrified the whole of the inner surface, which then runs into one single piece. But the vitrification does not penetrate above four or five lines into the blocks of lava, which are several feet thick, so that the part covered by the glass, being exposed to a somewhat less degree of heat, ends at length in being supercalcinated; its colour then becomes red, its hardness is lost, and when the kilns are taken down or repaired, it is easy to observe in the thickness of these lava blocks, the gradual action exerted by a fire so strong and so long continued. The reader will excuse this digression, which is by no means foreign to the subject.

KNOCK

MOUNTAIN OF BEN MORE, THREE MILES FROM AROS

*Lavas in tables, prisms, and irregular
masses, hard, sound, compact; internally
of a blackish-grey, externally of a dull
white; decomposed to the depth of four
or five lines, and exhibiting the primitive
elements of their composition; in some of
them are found kernels and specks of
white zeolite*

In mentioning the mountain of Ben More,
I said that it was covered with bushy
heather so thick as hardly to allow the lava
of which it is composed to be seen. But on
following several hollows excavated by the
waters which run down its sides, I could
only recognise one single kind of lava,
which is grey, hard, compact, with kernels
of zeolite. I carefully examined several of
these ravines from the bottom to near the top
of the mountain, and no where did I meet
with any other kind of lava. But, as the
way is one of the most difficult, I was able
to visit only the north side of this volcanic
peak. I therefore invite those naturalists

who may make the same journey, to attack
the south side of the mountain, to discover
whether the lavas there be equally homogeneous.

Knock is the name of the residence of
Mr Campbell; and to distinguish him from
other persons of the same clan, he is described
by the appellation of Campbell of Knock.

His house, situated on an eminence, at the
foot of Ben More, has a view on one side of
a charming valley, covered with flocks, and,
on the other, of a fine sea-loch, navigable,
full of fish, and visited by the herring at the
time of their migration.

A considerable clearing which he has
made in the midst of the lavas, to obtain a
little soil resulting from their decomposition,
has demanded an amount of labour which
nothing but the most obstinate perseverance,
supported by the hope of fertilizing and embellishing the place of his habitation, could
have been able to surmount.

This great undertaking has produced considerable accumulations of volcanic stones,
broken, set on edge, and cut in every
direction. The ground has been cleared
by gathering them into dry stone walls of

great extent and proportionate thickness. These numerous fences offer to the naturalist a fine field for observation.

The lavas are compact, of a black or rather deep grey colour, approaching to black; most frequently disposed in tables, sometimes in prisms, also in irregular masses. Their fracture presents a base, in appearance homogeneous, of a grain bright, equal and susceptible of a fine polish. But a peculiar alteration observable on their outer surface, and which penetrates a few lines into the interior, deserves all the attention of the naturalist, and makes lavas of this kind worthy of interest.

This alteration, effected by time, or rather by the different modifications induced by the air upon these lavas, has revealed their constituent materials. It may be regarded as a sort of natural dissection, which, by destroying certain parts, has exposed to view those which would otherwise have remained concealed, and which no chemical analysis could ever have recognised. This requires a more particular explanation; which I proceed to give, with a specimen in my hand, that those who may have occasion to observe similar lavas, common also in the ancient extinct volcanos of France, may be better able to correct my errors, if they shall think me mistaken, or make use with me of a means which may sometimes discover to what stone such and such a lava belonged before its fusion.*

The lavas in question, I repeat, present on a broken surface a hard, bright base, of a dark grey colour, approaching to black, in which the particles seem well amalgamated and homogeneous; without showing any difference between them, even with a lens.

If we proceed to the examination of their external parts, we find the surface granular, unequal and rough under the finger, and exhibiting crystals and plates of felspar, projecting specks of black schorl, often implanted in the felspar itself, both the one and the other being surrounded with small cavities, which isolate them, and which imply the destruction of the molecules amidst which the felspar and schorl were involved.

* [Those who may wish for further information regarding the rocks and structure of Ben More may consult the Editor's " Ancient Volcanoes of Great Britain," vol. ii.]

The white crystals of felspar are slightly touched with a reddish tint, that becomes a little deeper in the interstices into which it has been more difficult for rain-water to percolate, and to carry off the ochreous molecules produced by the decomposition of the iron.

The most experienced naturalist in the study of rocks, on seeing the decomposed surface of those here described, cannot avoid regarding them, at first view, as true granites. He only meets with embarrassment, when he examines their fracture and their interior grain, and above all when he presents to the magnet this unaltered part which attracts it as strongly as the basaltic lava richest in iron; whilst the external crust has no action upon it whatever.

It thence results, that the iron which forms one of the constituent principles of this lava, has had its nature completely changed, and in this alteration has carried with it the earthy molecules with which it was combined or which it held enchained.

This bond once broken up, the substances which were sheltered from decomposition, such as felspar, schorl, and some small portions of quartz have been exposed; the veil which hid them being once raised, we can readily recognise the organisation of these stones.

Their first origin appears, therefore, to be derived from a granite or a porphyric rock. The naturalist will more readily decide in favour of this latter rock, from the consideration that the base of the true porphyries is in general *petrosilex*, which, in spite of its hardness and whatever may be its colour, sometimes decomposes naturally in the open air, and is above all liable to be strongly acted upon by sulphurous acid vapours.

But to be assured that this lava owes its birth to a porphyric substance, with a base of petrosilex, nothing more is required than to fuse with the blow-pipe a small fragment taken from the soundest part, that is, from the part which has preserved its hardness and its black colour. It will then soon be seen that a white enamel is produced, a characteristic indication of petrosilex; whereas the lava with a base of *roche de corne**

* [The term *Corneus* was applied by Wallerius (1747) to various dark, heavy rocks and massive minerals, including hornblende and basalt. It was translated into French as *roche de corne*, but it is often impossible to say to what particular rock or mineral, as now defined, it was intended to apply.]

produces a fine enamel of a deep black colour. My learned friend, Deodat Dolomieu, has sufficiently established this distinction in his excellent memoirs.

Similar lavas are found at the foot of the Mont Mezinc in the Vivarais, near Le Puy, in Velai, on the Euganean mountains, and in the Ponza Islands.

One is always astonished on examining certain lavas, to find that the subterranean fires should have been able to melt into a stream, stones which now appear the hardest, and that without altering, so to say, their primitive organisation.*

LEDIRKILL

FROM AROS TO AUCHNACRAIG

White compact Lavas which have preserved their hardness

ON the way to Ledirkill, some hard compact and very white lavas are to be seen.

---

* [It may be gathered from this section of the book that the author looked upon these lavas, not as parts of the earth's constitution, which must for the present be regarded as original, but as the result of the fusion of portions of the solid terrestrial crust by "subterranean fires." This was the accepted doctrine of the Wernerian school, in regard to volcanoes and their products.]

They do not appear to have undergone any alteration, either spontaneously, or through gaseous emanations. Their base is fairly homogeneous; but the particles are a little scaly, and have some resemblance to those of a certain felspar. Their white colour seems to give no indication of iron. But one would be led into error by trusting to this deceptive appearance; for they have a very sensible action on the loadstone. There are ores of white spathic iron, rich in the metal, but not affording by their colour any sign of iron.

The white lavas of Ledirkill have some resemblance to tufas, with this difference that the latter are nowise magnetic, while the former yield no alum.

I therefore consider the lavas of Ledirkill as naturally white, and as probably owing their origin to stones of the nature of rocks with a base of petrosilex or with a base of massive felspar.*

* Déodat Dolomieu, who has so well observed the different causes which tend to decompose or bleach lavas, thinks as I do, that some of them are naturally white. " There are a number of lavas," says this learned mineralogist, " of a white or whitish colour, which have never been attacked by vapours, and which have not sustained the least alteration. This is proved by local circumstances, by the hardness, and the perfect preservation

AUCHNACRAIG

Beds of Limestone, between two banks of Sandstone, in the midst of the Lavas; Belemnites in the Limestone

ABOUT half a mile from Auchnacraig, and not far from the prostrate column which I have mentioned, and which the inhabitants regard as the work of Ossian, there is by the sea-side a cliff, upon which the waves beat with so much fury, that they have torn open the volcanic rock which has not been able to oppose an insurmountable obstacle to their advance.   By daily attacking this barrier for so many ages, the waves have brought into view a bed of limestone, formerly buried under a current of black basaltic lava, of which the whole coast is formed.   This bed, which averages about fifteen feet in breadth, is uncovered for a space of at least twenty toises in length at low water; it can be seen to lose itself in the mass of lavas which rise into hills as they recede from the coast.

of the felspar and the micas which they contain.   I could in-stance a great number of lavas which are naturally white; such are those of the Euganean Hills near Padua, named *granitello*, some lavas of Ætna, Germany, &c."   Mémoires sur les Iles Ponces, Paris, 1788, p. 37.

The limestone is grey, hard, and brittle, but is not very pure, being mixed with a small quantity of argillaceous earth; though fit for making lime. I found some belemnites in it, the largest of which were five inches in length, and an inch and a half in circumference towards the base.

This calcareous stratum does not adhere directly to the basaltic lava, being separated from it by two thin beds of quartzose sandstone with coarse grains, which are held together by a partly calcareous cement. It is to these that the lava adheres; and had not the sandstone been laid bare by the daily and violent action of the sea, one would never have suspected that under these enormous masses of basaltic lava, there exists a layer of calcareous material, itself intercalated between two beds of sandstone.*

* In the 160th and following pages of the *Minéralogie des Volcans*, I have mentioned some analogous, but much more remarkable appearances, when describing the mountain of Chamarelle in Vivarais, near Villeneuve de Bery, where beds of limestone and basaltic lavas may be seen alternating with each other, and where belemnites are found in the limestone, as in that of Auchnacraig. In that work I have stated my conjectures respecting the manner in which these different beds might have been formed at the remote period when these ancient volcanoes were active under the waters of the sea. [This notice of the occurrence of belemnites in the West Highlands appears to be

Grand natural basaltic Wall resembling an ancient Circus *

To the north of Auchnacraig, to the right hand on quitting the house, and about six hundred toises therefrom, and close by the sea, there may be seen a natural platform of semicircular shape, situated on an eminence about fifty feet above the level of the water, and entirely composed of black lavas of a basaltic nature. This little plain, which has a gentle slope, is bounded on the south by a perpendicular volcanic rock.

A vast detached wall surrounds a portion of the circle, formed by the basaltic rock which rises on the opposite side, the result is to present the appearance of a kind of antique circus, which fills one with astonishment at the first view, and gives to this singular place the aspect of a ruin, as extraordinary as it is picturesque.

The objects assume a new character of grandeur in proportion as we approach them, and the picture becomes more striking as we

the first on record. The science of Geology, however, was not then, nor for many years afterwards, sufficiently advanced to allow the full interest of the discovery to be appreciated.]

* [The author gives a view of this place in his *Essai de Géologie* (1809) Pl. 29.]

contemplate near at hand the height of the
wall and its astonishing regularity.

At first, we can hardly conceive how, or
from what motive, human beings should
have come to raise, in a place so remote and
so deserted, a monument presenting the very
image of a Roman circus.

The farther we advance, the more surprising does this kind of arena become. A
large angular gap in the midst of one of the
walls permits the eye to behold the interior
of this antique ruin. We feel a lively
curiosity intermixed with uncertainty, respecting the nature of the object before us.
Such at least were the impressions that I
experienced, as well as my companions, the
first time that we went to see this remarkable
place: even when quite close to it, we took
it to be a monument of art.

There is nothing here, however, but the
work of nature, and one of the most
extraordinary productions of the subterranean conflagrations; no less astonishing,
perhaps, in its kind, than that which gave
birth to the cave of Fingal.

I have said that a black basaltic rock, cut
down vertically and describing a natural arc

of a circle, forms the back of the circus. A
vast wall, perfectly upright, forms the remainder of the inclosure.

It was this wonderful wall that engrossed
our attention. It is eighty-nine feet long,
perfectly straight, and composed entirely of
prisms of black basalt, of equal length, and
placed horizontally above each other; that is,
all these prisms, which are in good preservation, and pretty equal, laid one upon
another, form the thickness of this wall which
is detached on both sides from all the surrounding rocks. Its facings are pretty even,
and though more than twenty-five feet
high, it stands erect without abutment or
buttress. Only at its northern end does it
rest against one of the fore-parts of the
volcanic rock which forms the circular back
of the amphitheatre.*

The gap in the middle of the wall is
fourteen feet four inches wide at bottom,
that is, even with the ground, and forty-two

* [This wall is a basalt dyke which has here been isolated by
the sea when the land stood some forty or fifty feet lower than
it now does. Had the author had time to continue his journey
along other parts of the shores of the West Highlands, he
would have met with many other striking examples of the same
feature, along the platform of the fifty-feet raised beach.]

feet towards the top. It forms a large obtuse angle, and gives a very picturesque appearance of ruin to the whole of the circus. This breach is probably the effect of an earthquake. I counted within the interior about forty prisms which seemed to have belonged to it, and thirty-nine outside. But these are nothing to what would be still necessary to fill it up; and it is not likely that any have been carried away by man. The sea is at present a hundred feet distant from the wall, and forty feet lower at ordinary tides. It is possible, however, that the waves may have opened the gap at some very remote period, and carried off the greater portion of the materials which are wanting. This conjecture will, perhaps, appear more probable than the first, when I shall have described other objects in the vicinity of the wall, which are fitted to throw some light on the theory of its formation. I shall shortly return to this subject.

Nothing can better convey an idea of this basaltic wall than the manner in which the wood for firing is arranged in the wood-yards of Paris. These pieces, as is well-known, are all of the same length, and are piled up horizontally above each other. I do not mean those enormous piles which overtop the houses, and form vast masses of wood; because, in these, the pieces are placed longitudinally and crosswise alternately. I refer to those kinds of wall, the thickness of which consists of the length of a single billet, and which are carried only to the height of ten or twelve feet, that the wood may be more at hand for daily sale.

I am obliged to use this trivial comparison, in order to make myself better understood. It is not easy to be clear, and at the same time to avoid wearying the reader with details too minute or imperfectly expressed, when it is necessary to describe objects which Nature seems to have produced in her capricious moments, to embarrass us with deviations of which she displays but few examples.

I am fully conscious of my inability to express what I saw, or what I felt, on beholding the volcanic circus in the vicinity of Auchnacraig. I therefore beg the most ample indulgence for what I have already said, and for what I have yet to say about it.

The height of the great wall is twenty-five feet ten inches. Its thickness is seven feet ten inches, and the prisms, of which it is composed, are consequently of the same length. These prisms are pentagonal, hexagonal and seven-sided; rarely quadrangular. The most common forms are the pentagonal and hexagonal. The rock of which they consist is black, hard, sound in fracture, and magnetic.

The first eight courses of the upper part of the wall are formed of entire prisms, in excellent preservation, and placed horizontally one above another without any adhesion; that is, they might be easily lifted away one after the other. But they lie so close together that there is no interspace between them save the lines of separation which define the prisms, so that the face of this singular wall imitates a kind of mosaic work.

The prisms which succeed the first eight courses are likewise unbroken; but they are cut transversely in some parts, either from the natural effect of contraction at the time of the lava's cooling, or from the weight of the incumbent mass at a period long subsequent to their formation.

The wall commences towards the west, where it abuts against the rock of lava which serves it as support. It then takes a south-east direction, and turning, stretches along towards the north-north-west, and towards the south-south-east. It is not of an equal height throughout. The most elevated part, which is also the best preserved, is twenty-five feet ten inches high, as I have stated already; the remainder is twenty-one feet seven inches. It is detached on both sides, and is, in all, eighty-nine feet in length, including the gap. The greatest diameter of the circus, which is rather of an oval than circular form, is seventy-six feet eight inches; and to bring all the measurements into one view, I may repeat, that the wall is a hundred and two feet distant from the sea, and stands on a site wholly occupied by lavas, raised forty feet above the level of the sea in mean tides.

It is doubtless very difficult to conceive how the lava, when flowing, could have formed a wall so high, of such regular construction, unconnected with any other mass, and composed entirely of diversely-sided prisms, placed horizontally by the side of each other, with such order and perfect symmetry, that the hands of the most able stone-cutter could never have arranged them with such dexterity.

This problem, however, which is certainly attended with great difficulties, finds, on the spot itself, some means of solution, arising from particular circumstances capable of conveying an idea of the manner in which this prismatic wall was formed.

For this purpose, it is necessary only to step about forty paces towards the south-south-east part of the circus, on the side next the sea, where two facts may be discovered which serve to explain this remarkable problem. I congratulate myself on having continued so long on the spot, and having so carefully traced all the surroundings of this singular volcanic monument; for otherwise this important observation would have escaped me.

Two large natural excavations in the lava itself may here be seen, one of which is twenty-two feet deep, sixteen feet broad, and a hundred and forty-six feet long, and the other eighty-five long, nineteen broad, and on an average twenty-one deep, these seem as if they had been designedly placed at no great distance from each other, to invite the observer to repair thither in order to learn how nature works in the construction of such walls.

Let the reader imagine to himself, for a moment, two streams of lava of a considerable thickness, which at the time of some great eruption have flowed parallel to each other, leaving an interval of several toises between them. The case is not without a precedent, at Etna, at the volcano of the Isle of Bourbon, and elsewhere. From these two streams there results a long and deep gallery, or a kind of covert-way, more or less straight, more or less circular or winding, according to local circumstances, and the obstacles which the streams might meet with on their way.

But in this case, granting that two currents of lava might indeed, as they approached each other, form such a gallery, still it may be asked, how could they follow a direction so equal and so parallel as to produce a channel nearly uniform throughout, and of which the interior facings present a perfectly even surface? I might reply, that this might occur, since instances of it are known, and I would add, that naturalists know very well, that in great eruptions, the lava does not flow along with the same liquidity as melted metals, but like a thick paste, and the air, by cooling the parts in contact with it, forces them to stand upright upon each other. This may be witnessed where the boiling lava presents vertical faces, creeping slowly along in that condition and yet continuing to a great distance. What is still more astonishing is, that such currents are sometimes seen to divide into two parts, like two branches of a river, on merely meeting with a body which they could easily overturn, such as a stone mound, or even a house. Sir William Hamilton has accurately observed and described this astonishing phenomenon, in his excellent descriptions of the eruptions of Vesuvius.

Other causes may further contribute to give regularity and smoothness to the interior facings of a gallery formed by two parallel currents of lava.

The volcano, for instance, may have been submarine, or only in the vicinity of the sea, where those at present in activity are

almost all situated. We have, then, only
to suppose that two currents of lava, flowing
along together, reach the water and advance
for a certain distance into it. The sudden
cooling, the resistance of the fluid, the thick
and deep mud which generally covers the sea-
bottom, or lastly a bank of shifting sand, or
other unknown causes, may give rise to what
so much astonishes us, namely, the parallelism
and equality of the interior facings.

It is of little importance to know the
exact and perfect theory of these works of
nature. It is sufficient that the fact exists,
and that it cannot be doubted, after examining the two large and deep galleries or
gullies mentioned above, which appear in
open view not far from the circus, and
which enable us to explain the formation of
the great wall. I have only to beg of the
reader a little further patience and indulgence for details, too long and tedious, into
which I am obliged to enter in order to
make myself intelligible upon a subject, dry
and difficult in itself, but well fitted to
interest those who are fond of this kind of
studies and observations.

The first of the two gullies was such as

strongly to excite our attention at first sight.
I have already said that it is eighty-five feet
long, nineteen broad, and twenty-one feet
of average depth. It is wholly uncovered.
There is no access to it, however, except at
one place, where, with a little address, and
with the help of some blocks of lava which
have fallen in and form a kind of steps, one
may descend to the bottom.

This long and deep excavation is the
effect simply of two currents proceeding in
the same direction, with an interval of
fifteen feet between them. The lava of
which they are composed is black, and of
the kind which I have denominated in the
Minéralogie des Volcans, "gravelly lava";
that is, which has little cohesion and falls
naturally into gravelly fragments, or lumps
of different sizes, having a general tendency
to separate in that manner, particularly in
the parts exposed to the air and to alternations of dryness and humidity.

Matters being in this state, and the
channel or gallery being formed, it then
served as a mould to a current of basaltic
lava, compact, homogeneous, and of great
solidity, which subsequently flowed into it,

and thus gave rise to a wall something similar to the cased walls of the Romans.*

As the current of basaltic lava would pour along the channel in a boiling state, its sides, that is, the parts in contact with the walls of the gully, must have necessarily been the first to cool. The caloric thus escaping by the sides, and the lava shrinking into smaller volume, prismatic contraction of the material must necessarily have taken place; the loss of heat and of gaseous emanations compelling it to shrink, and giving rise to a contraction whereby the moulded wall was divided into horizontal, variously-sided prisms naturally superposed upon each other.

The outer walls, which served as the sides of the mould, and which only consist of a gravelly lava, have merely to be denuded by the waters, either gradually during a lapse of time, or by some extraordinary movement of the sea. It would then be obvious that the wall, built of more solid materials, being thus stripped of its mould, would appear to have been erected as by

* [That is, walls of concrete which was filled in between two parallel palings of close-set wooden planks. The gullies here described were obviously dykes which may have been more or less eroded, together with the fissure walls on either side.]

miracle, and to have risen out of the earth like a piece of theatrical scenery.

Now this is precisely what has happened here, at least to such a degree as to leave no doubt of the fact : for, in the middle of the gully of which I have spoken, we see a perpendicular wall,* three and a half feet thick, and eight high, completely divested of all other lava, quite isolated, and entirely formed of prismatic columns laid horizontally above each other, but preserving a certain mutual cohesion, which has prevented them from falling down, and enabled them to resist the action of time and the elements, which they could not otherwise have withstood.

I never tired of admiring this wall, round which one can walk with ease; for the whole width of the gully being only nineteen feet, and the prismatic wall just four and a half, this striking mural rock stands nearly in the midst of an empty space of fourteen feet six inches, with seven feet three inches of that space on each side of it.

The vacant ground on either side of the wall was probably once filled up with the

* [Doubtless the dyke, so far as it has survived denudation.]

same gravelly lava of which the sides of the gallery consist. The sea which, during stormy weather and high tides, rushes noisily into the gully by an aperture that communicates with it, will have carried off the gravelly lava from the space in the middle of which the wall was incased.

It is probable that with time, and with the help of rains, hoar-frosts, and the sea, which are continually at work upon the gravelly lava of the gully, the wall will one day be found entirely stripped of any surrounding rock on either side, without the survival of the least vestige of the primitive mould to which it owes its formation.

It only remains for me to say a word regarding its present elevation, which is no more than eight feet, whilst that of the excavation in which it stands is twenty-one.

I reflected upon this fact on the spot, and I think I shall be able, partly at least, to account for it by saying, that the wall was once presumably higher, but that the upper courses, consisting of prisms which did not adhere to each other, must have been swept away by the sea.

This opinion gains some strength, from

an examination of the second gully, which
is at a small distance from the first, and on
which I shall dwell only for a moment.

This gully, much larger than the other,
is a hundred and forty-six feet long, twenty-
two feet deep, and sixteen feet broad. It
may be regarded, in one sense, at least, as
the reverse of the former. The two parallel
currents, which have served to form it, consist of black, compact, very hard lava, in a
single mass, which has resisted all the injuries
of the weather, the action of the air, and the
highest tides.

A stream [dyke] of compact, homogeneous
lava, has also found its way into this great
gully of which it occupies the whole length.
But the basaltic lava of this secondary current
had a paste so uniform and so well amalgamated, with so strong a tendency to divide
into regular and perfect prisms, that lying
horizontally upon each other, they had neither
bond nor cohesion, as far as can be judged
from a small portion of the wall which remains towards the beginning of the gully,
which the sea has not yet reached.

The prisms of this remnant are truly
astonishing from their excellent state of pre-

servation and the complete regularity of their form. They seem as if they had been placed there with all the care and art of human hands: so very wonderful is the symmetry and perfection of their arrangement. There is not one of these prisms, picked up at random, which would not be an ornament in a cabinet of natural history.

Their want of cohesion has been the cause of their gradual demolition; for the waves meeting the most obstinate resistance from the sides of the gully, which are of impregnable solidity, directed all their fury against the prisms, which they easily displaced and dragged off, finally consigning them to the bosom of the deep. Thus has this prismatic wall been almost entirely demolished, whilst the mould which helped to form it remains nearly unimpaired.

Such is the manner in which volcanos, so generally agents of destruction, can create, or rather imitate, by a succession of accidental circumstances, works which man cannot produce save with much labour and a series of long and difficult means and combinations.*

* I made these observations on the spot in the month of October, 1784. Déodat Dolomieu, three years afterwards,

From the preceding observations there is every appearance that the grand wall which forms the volcanic circus of Auchnacraig has had no other origin. But as the platform on which it stands is forty feet above the present level of the sea, and the wall itself is still twenty-five feet ten inches high, we must believe that the sea has fallen sixty-five feet ten inches in this district, unless we suppose that the coast has been elevated by the incalculable efforts of some vast subterranean explosion.

that is, in the month of July, 1787, on visiting the Ponza Isles discovered a similar wall, but with much smaller prisms. As the comparison may be interesting to naturalists, I shall make use of the language of my friend: "The small basalts are very numerous in the Ponza Isles. They are found in a multitude of places, but principally in the cliffs of Chiardiluna, to the left of the subterranean gallery. There are thousands of them on both sides of the small bay of Ste Marie, especially on the mountain in the rear of the houses. These small prismatic columns naturally come asunder and fall into the sea. Some of them are perfectly regular, and all the variety of forms of which they are susceptible are found among them. They are heaped together in different ways, but more frequently piled horizontally on each other, and rising above the ground in the form of walls which perfectly resemble those in the ancient fabrics called *opus reticulatum*. Several rows or walls made of prisms nearly a foot long, rise one behind another."

Dolomieu entertains the same opinion with me respecting the theory of these walls; he regards them as having been formed by inclosure within the fissures formed in the midst of the lavas. *Memoire sur les Isles Ponces*, by Déodat Dolomieu, Paris 1788, in 8vo, 98 and following pages.

Mr Anderson, who, at the same time as myself, travelled by command of the English government, with a view to the fisheries, through several of the Hebrides, told me, that he saw in the isle of Islay, a volcanic wall of the same kind with that of Auchnacraig, respecting which I had given him some details. He informed me, that the wall begins on the west side of the island at a place called *Cove*; that it describes a diagonal line three hundred paces long; that it is at least fifty feet high by four feet thick; that one-half of it stands out of the water, while the rest runs out into the sea, where it forms a most extraordinary jetty, which at first sight would seem to be a work of human design and construction.*

---

* [This long and ingenious explanation of the origin of these walls may conceivably now and then apply to cases in volcanic districts. But there cannot be any doubt that it is generally untenable. The walls of basalt, known now as Dykes, which can be counted by thousands along the whole west coast of Scotland, and which often form such striking features in the scenery, are certainly due not to the flowing of superficial lavas into moulds, but to the uprise of molten material from the earth's interior in open fissures of the terrestrial crust. These fissures, which traverse indiscriminately not only volcanic rocks but sedimentary formations of many various kinds and different ages, are in large measure vertical. The igneous material cooled and consolidated in them as the concrete of the Roman builders hardened within

the wooden hoardings between which it was poured.   In most cases the substance of the dykes is more durable than the encasing rock, which is consequently worn away by the action of the various denuding agents, so as to leave the igneous rock standing up as a wall.   Where, however, the igneous material is less durable than that through which it has risen, it is hollowed out at the surface, so as to expose once more the vertical walls of the original gaping fissure.

The opinion that the material of dykes has been filled in from above was long prevalent.   But it was shown to be erroneous by Sir James Hall from observations made by him in 1785, during a visit to the old crater of Monte Somma. He inferred, and his inference has been amply confirmed, that the dykes in that famous crater-wall must have been filled in by the ascent of molten material from below into cracks and fissures of the cone.   *Trans. Roy. Soc. Edin.*, vol. v. (1798), p. 71.

CHAPTER VIII

The Isle of Kerrera

THE isles of Mull and Kerrera are separated by a narrow channel which may be crossed from Auchnacraig in less than four hours. Kerrera almost touches the mainland, by a point which runs out towards Oban; for the strait on that side is in some places not above fifty toises broad. I walked across Kerrera diagonally in order to get to a small ferry-boat, at the end of the island.

A part of Kerrera is volcanic. Especially on the coast of the side fronting Mull, there are accumulations of compact lavas disposed in mass and in large streams. This basaltic lava sometimes takes the form of prisms, which are not very regular, at least in the places which I had an opportunity of examining. I also found some rocks of micaceous schist of a whitish colour, and others which were greenish with a fibrous

texture. These schists or *gneiss* are composed
of quartz, steatite and small scales of mica.

Near the rocks of mica-schist there is
found common slate of a deep grey colour,
approaching to black, the first beds of
which are almost even with the ground.
Quarries might easily be opened here with
great advantage to the country; the slate
would even become an object of commerce.
In this rock are found some brilliant pyrites,
crystallized in cubes.

Such were the objects which engaged my
attention in the isle of Kerrera, where one
sees pasture ground, some cultivated land
bearing barley or oats, with huts scattered to
the right and left, but few in number.

I remained only four hours on the island
which I crossed in its greatest length. I
found at its further end the ferry-boat above
alluded to; a small skiff managed by a
single man. I was tired and made the
boatman go straight to Oban, where I
landed in less than an hour, and where I
found Count Andreani with our carriages,
ready to start next day. In the evening, we
made every necessary preparation for com-
mencing our journey by day-break.

CHAPTER IX

Departure from Oban.—Dalmally.—Tyn-
drum. — Lead Ore. — Killin. — River-
mussels containing Pearls.—Description
of these Pearls and their origin

WE left Oban for Dalmally on the 7th of
October, at six in the morning.
The distance is about twenty-four miles,
along a stony road cut by ravines. We
arrived at the place of our destination about
seven in the evening.

Bonawe, of which I have already spoken,
is half-way; it is a small hamlet, built at
the union of a branch of Loch Awe, with
Loch Etive, which has sufficient water for
small vessels, and where salmon abounds.

We visited an iron foundry, at a small
distance from Bonawe. It stands in a
charming situation, embellished around
with woods, verdure and well-cultivated
land. A beautiful avenue leading to the
Loch, which had vessels floating on its

surface, made the scene all the more enchanting. This delightful spot formed a strong contrast with the barren and reddish mountains of porphyry, and the rocky debris piled up in irregular heaps all around.

We were agreeably surprised to find an establishment of this kind in so retired a part of Scotland, where the people have but the feeblest idea of cultivation or of art; but we learnt that it belonged to an English company, which had been induced to erect works at this place in consequence of the abundance of wood and water, and its proximity to the sea.

We waited upon the manager of the works, who received us very politely, and shewed us some iron of fine quality made here. When I expressed my astonishment regarding the ore of which I had not seen either indication or the slightest vestige all the way from Oban, he replied, that I was right, for that the ore used in this foundry was brought in vessels from Cumberland; he then shewed me some heaps of a red hematite, partly decomposed, of excellent quality and rich in iron.

This establishment appeared to be con--

ducted with as much skill as economy; but
the woods were beginning to be worked
out; they were not extensive enough, nor
of a quick enough growth to allow of
regular cuttings. It is therefore to be feared
that this foundry cannot be carried on much
longer.

On arriving at Dalmally I had the
pleasure of seeing our good friend, Patrick
Fraser, who supped and passed the evening
with us. He informed me of his new
researches in the poetry of Ossian, of which
he had recovered a few more fragments, in
the different excursions made by him for that
purpose among the inhabitants of the higher
mountains and little-frequented places of this
part of Scotland. He had also enriched his
collection with other poems made by the
bards of the country, but of much more
modern date. This worthy man, passionately fond of literature, and of gentle and
modest nature, is unfortunately like an exile
in the midst of these barren and melancholy
mountains, where, to exist, he must perform
the functions of a schoolmaster. I earnestly
wish that his condition may be ameliorated.
The Society, established at Edinburgh, for

the investigation of objects relating to the antiquity, science and literature of ancient Scotland would reap much advantage from the knowledge and activity of Patrick Fraser, who has the advantage of being perfectly master of the ancient language of the country, which has absolutely no relation to English.

Patrick Fraser begged that I would send him some French books which he wanted; and, on my return to Paris, I shall hasten to pay to him this small testimony of my esteem for his talents, and respect for his moral qualities.* "I can only in return," said he, "give you my address, and offer you my poor services in this country." I here with pleasure transcribe this address in English, that those who may feel an interest in becoming acquainted with this affable and amiable man may know where he lives. It was exactly as follows: "*Patrick Fraser, Schoolmaster, of Glenorchy, by Inverary, N. B. by London.*"

* I have sent him such books as I thought would be agreeable to him; but the distance and the difficulty of communication to this distant part of Scotland, have, doubtless, prevented me from hearing from him; it is even possible that he has neither received my letter nor my packet.

We slept at Dalmally, and next morning
took the road to Tyndrum; the journey
was only twelve miles, but we wished to
arrive there early, to have an opportunity
of examining a lead-mine of which we had
heard.

The valley of Glenlochy, through which
we passed, is pleasant in some places where
skirted with hills which are covered with
numerous flocks of sheep; but the mountains close in too much as one advances.
The soil becomes marshy and sterile; the
peat which is laid bare everywhere, gives a
black colour to the face of the country, and
a tint almost as sombre to the mind.

The hamlet of Tyndrum consists of only
a few houses, almost all detached; it stands
upon a low marshy piece of ground; a
humid and unwholesome vapour renders the
site very disagreeable.

The place where the lead-ore is found is
not far from the inn in actual distance, but
a good deal in height, for the galleries have
to be sought on a pretty high mountain
difficult of access. They are cut through a
grey micaceous, schistose rock, with much
white quartz: the vein of lead is found in a

vein of the latter substance. The ore is usually accompanied by pyrites or blende, and is fairly abundant. It is sometimes covered with pretty crystals of calcareous spar. The galleries in general are very badly kept up, and the works are negligently managed.

The pieces of ore are sorted out, broken with hammers, and then washed to separate the ore from foreign substances. Thus prepared, the ore is sent to a foundry in the valley at the bottom of the mountain and cast into the melting furnace : charcoal and turf being used, but I do not know in what proportion, because the foundry was not in working order at this time, on account of some repairs to the furnace. Besides, the English, as well as the Dutch, are very reserved in explaining their processes, even in the most simple arts, which they always carry on with a kind of mystery. In France, on the other hand, at all the factories, they are in general very complaisant in giving information on such matters as a visitor may wish to know about.

I observed under large sheds considerable piles of peats, and near them a heap of pit-coal. From this I presumed that the peat is used with a mixture of a fourth or fifth part of coal. The latter article must be carefully husbanded, on account of the distance of the pits, and the cost of land-carriage.

I should much like, for the benefit of our manufactories where wood begins to fail, that a similar mixture of peat and coal should be employed, where they can be procured. As I should have been glad to support my recommendation with an example, I begged of one of the superintendants of the works to inform me in what proportions the peat and coal were used here; but he turned a deaf ear to my request, and broke off the conversation to speak to me about something else.

It is very easy, however, to make experiments upon the subject, and they would be attended with success; particularly if made by persons well acquainted with the quality of the peat and turf which are to be used.

It appeared that the lead-mines of Tyndrum have formerly been more productive and valuable.

I left this place for Killin, by a road

as dismal as it was monotonous. I doubt whether another such exists. It is made upon a bottom of spongy turf, which permits the water to filter easily through its elastic and moving substance, without becoming marshy, for carriages pass over it easily enough.

But what makes this route oppressively tiresome is, that it goes on thus for several leagues, between two ridges not far apart, covered with a black turf, on which nothing grows but short heather and some yellowish mosses, that distil the water, drop by drop, on all sides.

The mind soon takes on some of the same gloomy hue, and grows more and more melancholy as one proceeds; but the scene suddenly changes when one reaches the end of this kind of sombre gallery, the horizon then expands, and the fine valley of Glen Dochart opens into view.

Here limpid and copious streams, teeming with fish, glide in serpentine meanders, through the most smiling verdure, and form islets shaded with large trees. Charming rustic habitations now appear, together with numerous herds of cattle and sheep, while the young shepherds and shepherdesses who tend them make the air resound with their songs, and animate the delightful scene with their dances.

This day we made twenty-four miles pretty briskly, and reached Killin before night.

Killin, though supposed to be a town, is in fact only a hamlet, consisting of a few scattered houses at the end of Loch Tay. The inn is plain, but tolerably good, and the landlord is a very civil man. We saw over the chimney-piece of a small parlour several native birds, which he had himself carefully stuffed with straw; among them was a white wood-cock, which William Thornton purchased, as he also did some heath-cocks.

We were about to sit down to table, when I was surprised to hear myself called by my name; the stranger who pronounced it asked to speak with me. I saw from his appearance and language that he was a Frenchman; his face seemed not unknown to me and I told him that I thought I had seen him in Paris, but that I could not at the moment recollect who it was that I had the

honour of addressing. "I am Bombelles,"
said he; "I travel, like yourself, for pleasure
and instruction. I am going to Port Patrick,
to embark for Ireland." It was from one
of our servants that he heard of my being
in the inn, where he had just arrived himself
with horses and one of Lord Breadalbane's
carriages, with whom he had been spending
some days.

I had never had any connexion with M.
de Bombelles. But two Frenchmen who
meet at the far end of Scotland soon make
acquaintance; and we had besides several
common friends. From the career which
M. de Bombelles followed, from numerous
military and other charts which he had
along with him, I judged that diplomacy
and politics were more to his taste than the
natural sciences and the arts, and that he
probably had some particular mission, very
foreign to the object of my studies. I
ought, however, to do justice to his talents
and activity and to say that he neglected
nothing which could be interesting to his
own country. This I had an opportunity of judging from some portions of a
well-written journal, which he communi-

cated to me at the time, and in which I
saw articles relating to rural economy and
commerce, and even to a curious physical
fact respecting a very extraordinary flux
and reflux of the waters of Loch Tay. I
had already heard this occurrence mentioned
at the Duke of Argyll's, at Inverary; and
M. de Bombelles, during his stay at Lord
Breadalbane's, being close to the lake,
sought the best information that could be
procured on the subject. I insert here the
note which he gave me, because as I was
to be at the place next day, it seemed to
direct my enquiries.

"Between the hours of eight and nine in
the morning of the 12th of September (1784)
the water of the eastern part of Loch Tay
retreated to the distance of more than three
hundred feet from its ordinary limits, and
the whole of that space, in which it was
generally three feet deep, was left quite dry.
The water on leaving it, flowed back
towards the west, and then meeting with
another wave, the shock raised both with
violence to a height of more than four feet,
and covered them with foam. The union of
the waters from opposite directions, formed

one large wave, which advancing southwards, always keeping a height of more than four feet above the level of the lake, remained in that state nearly ten minutes. Thereafter for an hour and a half this extraordinary kind of tide continued to subside until it disappeared. It is singular that during this phenomenon, the weather was perfectly fine, and the air entirely calm; no perceptible motion was observed at the other end of the lake. Two days after, the same appearance recurred; but one hour later, and not in so remarkable a degree."—*Note extracted from the Journal of M. de Bombelles, 9th of October* 1784.

M. de Bombelles[*] took the road to Invery; whilst I made a longer stay at Killin, in order to obtain as much information as possible about the pearl-fishery of the river Tay, which here falls into the lake, to which it gives name.[†]

The master of the inn, who obligingly complied with every thing that could gratify my

* This is the person who was shortly after appointed Ambassador to Portugal.

† [Strictly speaking, the River Tay does not flow into, but out of Loch Tay; the stream which enters at the upper end of the lake is the Dochert.]

curiosity, brought two fishermen whose particular employment was searching for pearls.

They led us to the river which runs in a
clear stream over sand and pebbles, and they
soon brought up several dozens of mussels,
from about three and a half to four inches
long, by a little more than two inches broad.
The colour of the shells on the outside is a
deep brown, inclining a little to green.   The
shell is thick, and of a fine mother-of-pearl
tint within, slightly tinged with rose colour.
I regarded this species as belonging to the
Mya pictorum of Linnæus, or at least as
very nearly resembling it.   [*Unio margariti-
ferus.   Linn.*]

Our fishers, in consideration of a pretty
large reward, which we promised them,
engaged to open these mussels in our presence upon the bank.   But they wished to
reserve the pearls, if any should be found,
in order to sell them to us separately.   To
this proposition we acceded.

Imagining from this that we should put
a higher value on those which might be
found in our presence, these cunning fellows
brought with them some pearls, which they
dexterously introduced into several of the

shells in opening them; and they appeared
to be well practiced in this sort of petty
trickery which, however, I detected in a
manner that greatly astonished and perplexed
them, and that may be mentioned here, as it
depended upon a fact that deserves attention
with regard to one of the causes which contribute to the formation of pearls.

I told them to open the mussels before my
fellow-travellers, whilst I went to amuse
myself with fishing some of them; but they
were to inform me when they discovered
any pearls. I was soon called and shewn a
tolerably fine pearl, perfectly round, and of
a good colour. I took the shell and the
pearl, and said that the latter had not been
found in the mussel shewn me. The fishers
assured me that it had, and appealed to the
testimony of my companions, who confirmed
their assertion. I assured my friends, however, that they were deceived, and begged
them to watch more narrowly the next
time. I retired a few steps, and a minute or
two after I heard them call out, " Here is
another." I went up, and on examining the
mussel said that the pearl had, for that time
also, been slipped into the shell. The pearl

was beautiful; and they asked a sum for it which was six times its value.

Our fishers were greatly astonished; for, as I was at some distance from them, they were sure that I could not have seen their motions; my fellow-travellers who attentively watched them, were themselves deceived, or at most entertained only a vague suspicion; so well skilled were these men in a trade which procured them a few additional shillings from travellers.

My art was so supernatural in their estimation, that they confessed the imposition, and frankly shewed us some other pearls which they had in reserve for the same purpose. They were very curious to learn my secret, which might save them the trouble of vainly opening a great number of shells, for they seldom found above one or two pearls in a week. But as they spoke no other language than the Erse, and knew not even a word of English, I could explain myself only by gestures; and, though what I had to tell them to notice was not difficult, I doubt if they quite understood me.

My secret consisted merely in examining attentively the outside of the mussels, and when neither of the valves showed any cavity or hole, but had a surface, smooth and free from callosities, I could pronounce, without fear of contradiction, that there was no pearl in such a shell. If, on the contrary, the shell had been pierced by auger-worms, or ploughed up by other worms, it would always be found to contain pearls more or less valuable, or, at least the first beginnings of pearls.

This observation, which I have found invariably true hitherto, was the result of some enquiries, in which I had been engaged a long time before, respecting the formation of this beautiful animal product. Buffon introduces the information, which I communicated to him upon this subject, in his article upon pearls (page 125, vol. iv. of the *Histoire Naturelle des Minéraux*.) At that time I found that the pearl-mussel is attacked by two classes of enemies. One is a very small auger-worm, which penetrates into the inside by the edge of the valves, by burrowing longitudinally between the different laminæ that compose the test. The small tunnel thus formed, after extending to the length of an inch or an inch and a half, doubles back in a line parallel to the first, and separated from it by a very thin partition of shelly matter. These two parallel lines mark the direction taken by the worm in going and returning; the points of entry and of exit being easily distinguishable by two small holes on the margin of the shell, generally beside the mouth.

The parallelism of the two passages may be demonstrated by introducing a pin into each orifice. At the inner extremity, however, there is a small circular portion, formed by the worm in turning round.

As these small channels or covert-ways are made in the part nearest to the mother-of-pearl, there is soon an extravasation of the pearly juice which produces a protuber-ance in those parts. The cylindrical bodies thus formed may be considered as elongated pearls, adhering to the nacreous layer of the shell. When several worms of this kind work at the same time near each other, and unite their labours, the result is a knob of pearl, if I may use the expression, with irregular protuberances, in which the ends of the passages are easily distinguishable.

Another sea-worm, much larger, belong-

ing to the family of multivalve shells, attacks
the pearl-shells in a much more injurious
manner. This worm is a pholas of the
species of sea-dates. I have in my cabinet
an oyster from the coast of Guinea, pierced
by these pholades, which have their hinge
in the form of a cross beak. The hole
which they bore is pear-shaped; and pearls
of this form are sometimes found, which
were in high estimation among the ancients,
and are at present very valuable in the East-
Indies: I shall give a more particular description of this rare species of pholades in
another work.

There are undoubtedly several other kinds
of worms which pierce the pearl-shell, and
form cavities more or less round, in which
the juice consolidates into pearls.

This observation which has probably been
made by others besides myself, will have
given to some persons concerned in the
pearl fishery, the idea of artificially perforat-
ing the mussels, and thus forcing them to
produce pearls. In London I saw some
pearl-shells brought from China, which had
undergone this operation: for the artificial
hole was closed with a piece of brass wire,

riveted to the outside of the shell like the head of a nail; the part of the wire which pierced the nacre, was covered with a well-shaped pearl, which seemed as if soldered by its end. It is probable that the Chinese, who have been so long skilled in the arts, did not make this discovery in our days. Their astonishing and multiplied industry teaches us that we are only a new people.

Broussonet, with whom I had a conversation upon this subject in London, at the house of Sir Joseph Banks, told me, that a person had assured him, that there was still another method of obtaining pearls. It consists in opening the shell with the greatest care in order not to injure the animal. A small portion of the inner surface is then scraped off, and in its place is inserted a spherical piece of mother-of-pearl about the size of a very small grain of lead-shot. This globule serves as a nucleus for the pearl; the nacreous juice envelopes it, and at the end of a certain time, produces a fine pearl. He said, that experiments of this nature had been tried in Finland, and repeated also elsewhere.

From these observations it may be inferred that the production of pearls depends, perhaps, as much and perhaps a good deal more upon an external and accidental cause, than upon a natural superabundance and extravasation of the nacreous humour.

There are very fine pearls in the river-mussels of Loch Tay, to judge from some which the fishers of Killin offered to sell us, at more than double the price of those which are in commerce. But these fine pearls are far from numerous there; on the contrary, a very great number are found, which the jewellers reject, but which, though hardly fit to serve as ornaments for the ladies, are yet very interesting for a cabinet of natural history, since they confirm the theory of which I have just spoken. Most of these pearls have little or no lustre; some are round, oval, or elongated and almost cylindrical; others are hemispherical like a button; some oblong ones have a contraction towards the middle which gives them the appearance of two joined together; and lastly there are examples of a somewhat conical shape. All are generally fairly large, and of a pale fawn or brown colour.

Pearls of this kind are so seldom met with, that they might at first be taken for little oculated agates prepared for being set, or rather for *bufonites*,* particularly those which have no mother-of-pearl. Their substance is very hard, and yields with great difficulty to the file.

The auger-worm, which occasions the formation of the pearls of Loch Tay, pierces the whole thickness of the shell, which is of considerable density and of a fallow brown colour. As the shelly juice oozes out from all parts of the orifice which the worm has formed, the pearl must necessarily partake of the quality and colour of the substance of the shell, from the outermost layer to that which lines and embellishes its inner surface. Hence rude, rough pearls arise, which have only a thin coating of mother-of-pearl upon one side. There are cases, however, in which the pearl is pure and brilliant throughout; where, probably, the extravasation has come from the nacreous coating only. This may have been occasioned by another kind of auger-worm

* [Bufonites (literally, toadstones) was the name given to certain forms of fossil teeth found in Mesozoic formations.]

which attacks the shell solely in that coating. It remains for those naturalists who would study further this interesting subject, to make more detailed researches, in an enquiry which has as yet been only outlined.

CHAPTER X

Kenmore.—Extraordinary Flux and Reflux of Loch Tay

WE now took leave of our pearl fishers as well as our landlord, who had obligingly given us all the information in his power, and took the road to Kenmore, along the left bank of Loch Tay, skirted on either side by granitic mountains, that enclose the lake and at the same time confine the landscape within a very narrow compass. The foot of these mountains is tolerably well cultivated; but the soil only yields oats, which are not usually reaped till towards the middle of October. These oats are very tall; they were only beginning to be cut at the time I passed; I measured several stalks, and found the shortest to be four feet high, and the longest five feet six inches.*

---

* I do not entirely agree with Knox, who visited this place after me, when he says, "that the banks on both sides are

The lake is fourteen miles long, by about a mile in mean breadth. I had no certain information of its depth.* It contains fish, and its waters are soft and clear.

The mountains nearest the lake and forming its enclosure, are composed of a schistose micaceous rock, intermixed with felspar and quartzose material, the latter appearing to be most abundant. In this rock I found a few ill-shaped garnets, of a coarse and opaque character.

On reaching the southern [north-eastern] end of Loch Tay, one finds on a pleasant spot a fairly commodious inn, several houses, a new-built church, and a bridge thrown over a small river which issues from the lake; the whole is surrounded with trees, which enliven this pretty landscape. The name of the place is Kenmore.

The traveller begins here to perceive that fruitful, populous, and finely diversified by the windings of the lake, and the various appearances of the mountains." The views upon both sides are too confined, and exhibit only a melancholy and monotonous aspect; and the few scattered patches of oats, present only the picture of an ungrateful soil.

* The same author, speaking of the depth of the lake, says, "that it varies from fifteen to a hundred fathoms;" this appears to me very extraordinary. [The deepest part of Loch Tay is 510 feet and the bottom there is 164 feet below the level of the sea.]

he is drawing near to a more open country, and that he is about to leave behind him the barren mountains of the north of Scotland. The air which he breathes, the cultivation, the people, every thing proclaims it; and this first shade of change fills his soul with one of those gentle pleasures, which I cannot better express, than by comparing it to that which one feels on the return of spring, though at this time we were at the end of autumn. But it may be said that all is winter, all is wild, dreary and sterile in the regions which I had just traversed.

My first care, on arriving at Kenmore, was to procure the most exact accounts respecting the flux and reflux of the lake, which happened on the 12th of the preceding September; for it was here that the phenomenon first took place, and could be observed with care.

The master of the hostelry to whom I applied, and who spoke English, delighted with the kind of reputation which this event had conferred on the place of his residence, conceived that I had come expressly from France to see a country, which, in his opinion, deserved so just a celebrity. On this account he received me most cordially, and I have pleasure in expressing here my grateful acknowledgment.

"I cannot," said he, "have the pleasure of explaining myself how, what you wish to be informed of, happened, because I was absent from home, on the first day of the lake's motion. But I can direct you to two men who saw, and followed, the whole thing, and who will shew on the spot how it took place. One of them, in particular, who is a lad of intelligence, has carefully observed all the facts; and you will have reason to be satisfied with what he will tell you. You may, however, examine both; I shall go and give them orders to accompany you, and to answer all your enquiries; for they are both my servants."

The one was called James Allan, the other John M'Kenzie. The latter, although the younger of the two, had much more intelligence, and a more sensible way of looking at things than his comrade, and was less disposed towards the marvellous.

M'Kenzie told me, that at nine of the morning on the 12th of September, the sky being clear and the weather calm, a peasant who was washing his hands in the lake near where the river issues, observed the water leaving the bank in a very perceptible manner, which obliged him to advance a few paces farther; but it still seemed to retreat from him. This surprised him so much, that he hastened to inform his neighbours of it; one of whom then told him, that at sun-rise, having heard a noise like that of a sudden gust of wind, he went to the window, when, to his great astonishment, he saw the water receding from its banks, as if driven before a strong breeze; but as the weather was perfectly calm, his astonishment knew no bounds.

M'Kenzie having told me these details which he had himself from the peasant, I requested to see this man. He was sent for, but he had gone to a village six miles off. Not having, therefore, heard him myself, the details which I give, cannot be regarded as positive facts, peasants being usually much inclined to the marvellous. The noise, which, according to his account, preceded the ebbing of the waters, appeared to me somewhat apocryphal. M'Kenzie was of the same opinion.

The latter, continuing his report, said that he was not told of this extraordinary movement of the lake until ten of the morning on which it first showed itself. He instantly went to the edge of the lake, where he remained more than an hour and a half, watching with the greatest attention what took place. During this interval, he quite distinctly saw the water ebb and flow ten times successively; and the same alternate motion continued for the whole of that day.

He communicated to me all these circumstances at the edge of the lake, on the very spot where he stood when making his observations. He pointed out a large stone pretty far in the water, as the limit of the retreat of the water.

As this part of the lake was only three feet deep, I had the space carefully measured between the stone and the water's edge, and found it to be one hundred and sixty French feet. But John M'Kenzie had the civility to inform me, that when the phenomenon happened, the water of the lake, being rather low, came eight feet short of its present margin. To prove this, he pointed to a

stake which he had then driven into the
ground, and which was now standing eight
feet back in the water. The exact measurement, therefore, of the space abandoned by
the water, was one hundred and fifty-two
feet. From this, we see that the account
given to M. de Bombelles at Lord Breadalbane's, in which this space was stated to be
three hundred feet, requires to be corrected.
It ought, indeed, to be observed, that as
M'Kenzie was not on the spot until an hour
after the water began to move, it is possible
that the first impulse, which he did not see,
might have been much stronger, than those
which followed. But as the alleged fact
has not been proved, whilst what fell under
M'Kenzie's observation is ascertained by precise measurement, it is better to adopt his
account.

The lake exhibited the same phenomenon
on the following day, and likewise on the
third day, but in a less frequent and regular
manner.

Nobody observed the water during the
night; so that nothing is known of what
then happened.

M'Kenzie told me that when the ebb

took place, the water receded from the bank without any shock, or hurried movement, but tranquilly, as far as the large stone, from which it returned in the same slow and gradual manner to the margin.

The inhabitants of Kenmore, whom I was in a position to consult, all agree with M'Kenzie in the following facts: 1st. On the fourth day, the movements of the lake only occurred at much wider intervals. 2dly. On the fifth, sixth and seventh days, there was absolutely no ebb of the waters. 3dly. On the eighth, the movement appeared for a few hours only; and this was the case during two whole weeks, with intervals of two or three days, entirely without move ment. 4thly. The movement abated gradually, and the lake resumed its former position. 5thly. During the whole of this time, there was no violent wind, nor the slightest shock of earthquake.

These are the facts most deserving of credit. I have thought it my duty to give them here, for the purpose of dispelling those additions of the marvellous, with which some English newspapers did not fail to embellish them. Having myself collected all this information on the spot, some confidence may be put in it. This is not a suitable place for entering into an examination of the causes which might have given rise to this wonderful flux and reflux in a lake where such a phenomenon had never occurred before. Similar phenomena have, indeed, taken place in other lakes. But we have not as yet a sufficient number of facts on this subject, and those which are already known have not been collected by persons sufficiently practised in the difficult art of observation, to enable us to form any satisfactory conjectures.*

We had scarcely left Kenmore on our way to Dunkeld, when we were agreeably surprised to find ourselves on a road bordered with superb Scottish and American pines, and other beautiful evergreens, kept in good order, disposed with taste, and which gave to this route a charm and a movement that betokened the nearness of some great habitation.

---

* [The oscillations of level in lakes, now known as *seiches*, have been carefully studied in recent years. In some cases oscillations may be caused by subterranean movements, but probably they are more usually the effect of atmospheric disturbances. On the Lake of Geneva, where they are not infrequent, they occasionally reach an amplitude of two metres.]

About a mile farther, we came to the
enclosure of a vast park, decorated with
plantations of various kinds, in the midst of
verdant lawns, and traversed through its
whole length by the river Tay, which is
crossed by two or three bridges of different
constructions. Numerous herds of deer were
feeding in this beautiful place; sheep, cows
of fine breeds, and horses of various kinds,
gave to the scene an air of wealth, usefulness,
and taste, which displayed the intelligence
and fortune of the proprietor. A vast pile
of building, partly in an antique and partly
in the modern style, closed the further end
of this magnificent landscape. It was the
residence of the Earl of Breadalbane.

I had heard so much good of this rich
landlord nobleman, whose chief occupation
is to spread industry and happiness around
him, that I was sorry I had not procured a
letter of introduction to him, with which
the Duke of Argyll would have certainly
obliged me. But it did not form part of my
original plan to pass through Tyndrum or
Kenmore; on the contrary, I meant to return by way of Inverary. On my second
arrival at Dalmally, I resolved to change the

order of my journey, with the view of
verifying the fact of the ebb and flow of
Loch Tay. I regretted the loss of this
opportunity of gaining information as to the
extensive operations in agriculture and rural
economy, undertaken with so much success
by the Earl of Breadalbane, and of making
the acquaintance of a man with so excellent
a reputation, and so useful to his country.

We dined at a very clean and commodious hostelry, about a mile from Dunkeld,
opposite to that charming little town, and
built on an eminence surrounded with
woods and rocks. A large ruinous ancient
gothic church gives to this place a very
picturesque appearance.

We reached Perth a little late in the
evening, by an extremely stony and most
fatiguing road.

CHAPTER XI

Perth, its Harbour and Manufactures.—
Mr M'Comie, Teacher of Mathematics;
Mr M'Gregor, Teacher of the French
Language at the Academy.—Volcanic
mountain of Kinnoul.—The Agates found
there

THE small town of Perth stands in a
very agreeable situation on the river
Tay, which the tide here enters and makes
navigable for small vessels.   It is in a pretty
flourishing condition, and contains a population of about twelve thousand souls.

The stone bridge over the Tay was
constructed by the same person who built
that of Blackfriars, at London.   It is very
well made, but rather narrow for its
length.

William Thornton had an acquaintance
at Perth, of the name of M'Comie, who was
teacher of mathematics in the college, which
bears here the title of academy.   We paid

him a visit, and he was so kind and complaisant to us as to be constantly with us during our stay at Perth, where we remained nearly a week. He was very useful to us, as well as one of his colleagues, Mr M'Gregor, teacher of the French language, who had received his first education at Paris, and who was kind enough also to favour us with his company, and to take us to several manufactories.

Before the reformation in England and Scotland, the town of Perth, where catholicism reigned in all its splendour, contained some considerable religious foundations, besides a number of churches. Of these, the greater part are in ruins, or converted into churches for the use of the presbyterian religion. In several of the streets are seen some fine gothic façades, which once belonged to cathedrals, chapter-houses, monasteries, and nunneries. These remains of monuments, consecrated to a worship formerly so flourishing in the three kingdoms, proclaim that religions, as well as governments, have their periods of instability and revolution, which continually succeed each other at longer or shorter intervals, but which, when the impulse is once given, no human power is able to arrest: so true is it, that in morals, as in physics, nothing is durable in this world.

Machines for carding and spinning cotton, had just begun to be introduced at Perth when I arrived there. We saw the first of them at the manufactory of an individual who had them made at Manchester. He found it impossible, however, to convey them out of that town except by night; so jealous are the manufacturers of Manchester of this happy invention of Arkwright, which has given such wide celebrity and immense advantages to the commerce of that town.

The most considerable manufactures of Perth are fine cloths, of thread and of flax; and some very excellent articles of this kind are made here. I saw a loom for weaving very large bed-sheets, in one piece, by means of shuttles fixed on small rollers. A pair of these sheets, made of very fine linen, costs from a hundred and fifty to a hundred and sixty livres of French money.

I purchased at a manufactory of table-linen a dozen small napkins and a tea-cloth.

They were of excellent quality, and cost me
four louis. I was glad to carry them to
France by way of models.

I was also shewn, with an air of mystery,
at the house of a rich manufacturer of fine
linen, an instrument as ingenious as useful
for ascertaining with the greatest precision
the fineness of texture of a cloth.

It consists of a kind of small microscope
of great simplicity, which, instead of a stage,
has a round hole, about three lines in
diameter. The glass or lens corresponds to
that circular aperture at the distance of the
focus. The instrument is placed upon the
cloth, the threads of which are so magnified
by the lens, that the observer can easily
count how many are contained within the
space of the hole. It is evident that the
greater the number, the finer is the fabric of
the stuff. It likewise shews whether the
thread is too flat. The artisan who has
been taught the use of the instrument, if he
should present a piece of cloth which he
charges as fine, has no valid excuse when it
is found to be of a coarser quality, and
when this is proved to him by making him
count the number of threads himself. The

weavers, therefore, by its means, have become
accustomed to great precision.

The wholesale dealers equally employ it
in their purchases; and this is why they do
not like that everybody should be acquainted
with it; because, with its assistance, they
can transact their business on a surer footing
than those who are obliged to depend upon
the naked eye. I brought one of these instruments to France, where they were soon
multiplied.

Volcanic Mountain of Kinnoul, in the vicinity of Perth

The desire of examining the Hill of
Kinnoul, was what principally determined
me to pass through the town of Perth, from
which it is only two miles and a quarter
distant. I was therefore able to make
frequent visits to it during three days when
I stayed at Perth.

The collection of lavas and agates which
I made there was large. I spent half a day
and a whole night in writing out their
labels and in gumming its label on each
specimen. The finest specimens were dupli-
cated, and in some cases triplicated, for the

purpose of distribution among the naturalists of my acquaintance. The whole filled a
large packing-case.*

Scarcely does one cross the bridge of
Perth, before coming upon lavas in sheets, in
amorphous masses, and in ill-shaped prisms.
These different currents are connected with
the eminence of the Hill of Kinnoul, the
base of which covers a considerable extent of
ground. Following the road along the Tay,
with the mountain on the left hand, for two
miles and a quarter, one arrives at a very
steep and almost perpendicular rock, nearly
six hundred feet high, and on the very edge
of the road. This is the point to be first
made for, because it is the richest in agates
and other productions worth collecting.

Though the mountain appears extremely
steep here, one may, notwithstanding, with
some precaution clamber up to its summit.

* This packing-case, altogether with my whole collection of
the products of Scotland and the Hebrides, which was in the
best order, was lost upon a sand-bank, near Dunkirk. The
vessel which carried them from Edinburgh went to the bottom,
but happily the crew were saved. I thus lost all the fruits of
a toilsome journey, except a small box of the most remarkable
articles, which I would not part with, but kept beside me in my
carriage. I had taken care, however, to copy into a register the
catalogue of all my collections.

But, for this purpose, it is necessary to have the help of a stout iron-shod stick; and the climber need not then hesitate to scale the craggy rocks. He will be rewarded by attacking the hill from this side, because he can there read its structure, as it were, for he is then in a position to observe the forms and various dispositions of the lava-currents. The following is a note of the different objects which I collected:—

Volcanic Mineralogy of Kinnoul

1. Black, fine-grained basalt with a homogeneous base, forming an extensive current, in contact with a stream of black porphyric lava, with a base of trap, and so disposed as to leave no doubt that the basaltic lava in this state derives its origin from the porphyric lava. The latter has preserved its crystals of felspar, which are small but well-defined, whilst the lava of the basaltic stream has lost its crystals, which are amalgamated and blended with the very base of the porphyry, whether by a fiercer access of fire, or by a long-continued action. On examining the basaltic lava with a lens, there can still be seen in some parts of it, small crystals which

are not entirely amalgamated with the lava; and this passage may be pretty well traced, by the aid of external characters. If small splinters of the porphyric lava are treated under the blow-pipe, they afford an enamel of a beautiful black colour; the basaltic lava yields a glass or enamel absolutely similar.

2. The same basaltic lava, divided into large prisms, rather irregular, though well defined. These prisms present nothing in their fracture but a quite homogeneous lava, without the least crystal of felspar.

3. Basaltic lava of a delicate green tint, very hard, sometimes sonorous when struck, disposed in a large stream. This greenish lava cuts transversely a stream of black compact lava. Its greenish colour is owing to a particular modification of the iron. I was well aquainted with the earth of Verona, which takes its origin from the very remarkable decomposition of a volcanic product; but I had never before seen a current of compact, hard, and sonorous basaltic lava, with this greenish colour.

4. A quadrangular prism, well defined, in excellent preservation, and of an agreeable delicate green colour. I found it among the

debris of a considerable mass of lava of the same colour, which had fallen from the top to the foot of the hill.

5. The same greenish basaltic lava in tabular form.

None of the green-coloured lavas were magnetic.

6. Compact porphyric lava, with a black ground, strewn with abundant crystals of white felspar, which have not undergone any alteration. This lava is strongly magnetic.

7. A quadrangular prism of blackish porphyric lava, magnetic, with a knob of flesh-coloured agate on one of its faces.

8. Porphyric lava, crumbling into detritus, and forming considerable streams. I have no doubt that if this gravelly lava, which is not very hard, were reduced to powder by the aid of stamping-mills, like those used in Holland for crushing the lavas or *trass* from the neighbourhood of Andernach, it would afford pozzuolan, an excellent cement, of great and indeed indispensable use for all hydraulic constructions.

9. Compact porphyric lava, with a ground of deep iron-grey, inclining to violet,

with globules of green steatite, kernels of variously coloured agate, as well as globules of white calcareous spar; disposed as a large stream.

10. Compact porphyric lava, magnetic, with kernels of white, and sometimes flesh-coloured calcareous spar, and globules of steatite of the finest green-colour.

11. Reddish-coloured, compact, porphyric lava, forming a coulee between two streams of basaltic lava of a delicate green tint, and adhering to them.

12. Black porphyric lava, magnetic, intersected with bands of red porphyric lava, resembling the antique red porphyry. This lava, wherein both the porphyries are united, is very remarkable.

13. A geode of agate, internally lined with brilliant crystals of violet-coloured quartz, with hexagonal pyramids, enclosed in compact porphyric lava, dark-brown inclining to violet, with some kernels of white calcareous spar, and globules of agate and green steatite.

14. A geode of bright red agate, having in its interior a brilliant crystallization of white quartz of the greatest purity. This

geode lies in a black porphyric magnetic lava.

15. Oculated agate of a delicate rose colour, surrounded with dark-brown compact porphyric lava, intermixed with globules of green steatite. This specimen is very pleasing to the eye.

16. Red striped agate, inclosed in black strongly magnetic porphyric lava.

17. Semi-transparent agate of the most vivid red, in a porphyric lava inclining to violet, with lumps of white calcareous spar and globules of a delicate green-coloured steatite.

18. A geode with a crust of chalcedonic agate of a bluish tint, internally lined with crystals of brilliant quartz. Inside the crystals, particles of black lava can be seen, which have been caught up during the process of crystallization; from which it is beyond doubt that the formation of the geodes was later than that of the lava.

19. A lump of sparkling, white calcareous spar, disposed in rhomboidal plates, enclosed in a thin envelope of steatite of a fine green colour. The whole is enclosed in a black compact magnetic lava, more nearly resembling basalt than porphyry.

20. A lump of green steatite, within a thin crust of white calcareous spar, and imbedded in a porphyric lava, of a brown colour, inclining to violet. This fragment is the reverse of the preceding.

Such are the most interesting specimens which I gathered on the Hill of Kinnoul. I have no doubt that a longer stay would have enabled me to augment my collection. But others will be able to complete what I have here merely sketched. I had obtained neither direction nor guide for this hill. They had not even the least idea at Perth that it was volcanic. They only knew that some lapidaries from Edinburgh visited it from time to time in quest of agates, which they polished and of which they made a petty traffic.*

* [It would have been interesting, had Faujas collection reached Paris, to compare his lithological names and descriptions with those which modern petrographical research has given to the same rocks. The most important part of his narrative is his recognition of the true volcanic nature of the rocks of Kinnoul Hill, which at that time were classed with other similar rocks, as of aqueous origin. He was thus the first geologist to lead the way in the unravelling of the volcanic history of the Old Red Sandstone of Central Scotland.]

CHAPTER XII

St Andrews.—University.—Library.—Old Churches.—Natural History

LEAVING Perth for St Andrews, and passing through the small town of Cupar in Fife, where we changed horses, we made the journey in seven hours. All the hills on the road are formed of blackish gravelly lava and basalt.

We had letters of recommendation to Mr George Hill,* professor of Greek, and Mr Charles Wilson, professor of Hebrew, in the University of St Andrews. We saw these gentlemen next day, and they were both most eager to oblige us, and to procure us such information as could gratify our tastes and curiosity.

University

This University recommends itself to the notice of the traveller by the name of the

* [George Hill, D.D. (1750-1819) Principal of St Mary's College, St Andrews (1791-1819) joint-professor of Greek (1772-88) Moderator of General Assembly (1789).]

celebrated Buchanan,* who was professor of philosophy there.

There were formerly two colleges which are now consolidated into one.†   There was a professor of the Latin language in each of the colleges; one of the professorships is now suppressed, and a chair of Civil History has been created in its stead.   The Greek professorship is also of recent erection.

The revenues of the professors, who are thirteen in number, amount together to fifteen hundred pounds sterling, which gives a fixed salary of nearly three thousand French livres for each place.

The names of the professors are as follows:

Joseph Maccormick [1733-1799] Principal [of the United College.]
James Flint, professor of Medicine.
John Cook, Moral Philosophy.
George Forrest, Natural Philosophy.

* [George Buchanan (1506-1582) studied and taught in various colleges in France; returned to Scotland and became Principal of St Leonard's College, St Andrews (1566-1570); tutor to James VI: famous for his Latin writings, especially his poems and his Scottish history.]

† [There were formerly three colleges, namely, St Salvator's, St Leonard's, and St Mary's; the two former were united in 1747.   St Mary's College is now the Faculty of Theology.]

Nicolas Vilant, Mathematics.

John Hunter [1745-1837] the Latin
language.

George Hill, the Greek language.

W. Barron, Logic.

Hugh Cleghorn, Civil History.

Dr J. Gillespie, }
Dr Henry Spence,} Theology.

William Brown, Church History.

Ch. Wilson, the Hebrew language.

Library

The [University] library is open to the
public for seven months in the year, during
which it may be entered every day at stated
hours. There are likewise some other days
of the year upon which it is open. The
revenues appropriated to the maintenance of
this establishment arise from an old ecclesi-
astical tithe which was seized upon by the
crown and afterwards assigned to this library.
The annual income from this foundation
does not amount to more than thirty-six
pounds sterling, a sum not nearly adequate
for the most urgent current expenses. But
some casual emoluments from the admission
of graduates increases the total revenue of

the library to the sum of an hundred and
fifty pounds sterling. The number of books
is not more than eleven or twelve thousand,
almost all modern, except several bibles and
some devotional books, among which there
are only common things. I only saw a
manuscript somewhat interesting on account
of its excellent preservation : it is a Saint
Augustin, on vellum, of the thirteenth
century. There is also shown as an object
of curiosity, an Egyptian mummy in very
bad condition, without even its ancient case,
and appearing to me to be one of those
which the Arabs fabricate out of patches
and fragments, for the purpose of selling
them to such as are unable to detect the
imposition.

Ancient Catholic Churches

This town, during the time of the catholic
religion, enjoyed archiepiscopal pre-eminence.
The famous Cardinal Beaton was its archbishop. Large and superb churches pro-
claimed the opulence of their founders, and
the generous sacrifices of a people strongly
attached to their mode of worship. The
ruins of all these monuments, of which

there are still some fine remains, give the
town an air of antiquity, which forms a
singular contrast with the simplicity, the
modesty, and I had almost said, the poverty
of the greater part of its present habitations.

The church of the Second College, as it is
called, which is still standing, appears to be
very ancient. The steeple is a high tower,
of square form, and of a good and solid
construction. The church is spacious, and
in the gothic style; it is dedicated to the
Presbyterian worship, and contains the tomb,
now partly in ruins, of an archbishop who
founded the University of this city. This
monument is built in the wall with stone
of a very common kind, and exhibits
nothing remarkable. On an occasion of
making some repairs, there was discovered
within it a church mace, of gilt copper, four
feet long.* This ensign of dignity, which
I was permitted to examine closely, is
charged with gothic ornaments finely executed, but in bad taste. It is covered with
small steeples, and niches in which are

* [The monument here referred to is the tomb of Bishop
Kennedy, within which were found six maces; three of them
were distributed among the other Scotch universities, and the
remaining three are preserved in that of St Andrews.]

cowled and praying monks, also winged
angels gesticulating in pulpits placed at the
angles. Gothic medallions suspended all
round serve as ornament; and the whole
is surmounted with a figure of Christ on
foot, standing upright in a pyramidal niche.
This work, to judge by its style, may be
at most two hundred and sixty or three
hundred years old; it can only serve to
give us an idea of the arts and of the bad
taste of these times.

I likewise visited another church, which,
from an inscription on one of its doors, appears to have been built in the year 1112.
In this church we saw a large mausoleum of
white marble, representing an archbishop, as
large as life, kneeling, and an angel placing
a martyr's crown on his head. A large bas-
relief, at the foot of this monument, exhibits
the same archbishop attacked by some men,
in the garb of Scottish Highlanders, who are
killing him. A young girl in tears, detained
by some other persons, near a coach, which
they have stopped, struggles to go to the
assistance of the archbishop, in whom she
seems to have the most tender interest. De-
spair is depicted in her gestures and her face.

This scene instantly brought to my remembrance the sinister event, when on the 29th of May, 1546, Cardinal Beaton was killed by Norman Lesly, eldest son of the Earl of Rothes, accompanied with fifteen conspirators disguised as Scottish Highlanders. This prince of the church was doubtless a man of great talents, but ambitious, insolent, a cruel enemy of the Reformers, and he had the abominable cruelty to have the unfortunate George Wishart burnt alive.

I was astonished to see a monument of this kind in a church now in use by these same Reformers who have held Beaton in such abhorrence.* But my astonishment soon ceased on learning that this monument, the sculpture of which was executed in Holland, had been erected by the relations

* [The author's reading in Scottish history did not extend, perhaps, much later than the time of Mary Stuart. He did not realise that in the tragic political annals of the country there could have been two Archbishops successively murdered at St Andrews. The monument which he saw in the parish church did not commemorate Beaton, but James Sharpe who, at first a presbyterian minister, afterwards espoused the cause of episcopacy in Scotland, and ultimately became Archbishop of St Andrews in the episcopal hierarchy set up under Charles II. He was believed by the presbyterians to have betrayed their cause, and was cruelly murdered by a fanatic party of them, in presence of his daughter, near St Andrews in 1679.]

of the archbishop a long time after his death, and that they had set apart a certain yearly sum for keeping it in repair; so that in order to obtain this sum, the mausoleum must necessarily be allowed to exist as a proof. But it also comes about that no repairs are made on the monument, though it begins to be in great need of them; and the income is probably applied to the use of the church. No part of it, however, will be demolished so long as the allowance is paid: an evident proof that every where, and in every case, gold conciliates men and even very often reconciles the most opposite opinions.

It would appear that the relations of Cardinal Beaton had no wish to conceal that the holy archbishop was a father, since his daughter is represented in tears, with her arms stretched towards her father, and held back by two of the conspirators, at the moment when the others accomplish the murder. But the solemn Robertson * informs us, in his *History of Scotland*, that the prelate openly acknowledged this daughter. " Cardinal Beaton," says he, " with the same

* [*Postea*, p. 225]

public pomp, which is due to a legitimate
child, celebrated the marriage of his natural
daughter with the Earl of Crawfurd's son;"
and in a note he says, "the marriage articles,
subscribed with his own hand, in which he
calls her *my daughter*, are "still extant,"
(Vol. i. p. 88, of the 8vo edition.)

The façade of the church of St Leonard,
though gothic, presents an elegance and
an air of grandeur which is impressive.
There was formerly here a college which
has been united to the University. Johnson
in his *Journey to the Western Islands of
Scotland*, complains, that he was always
by some civil excuse hindered from entering it, and he says that in fact it had
been turned into a green-house. I was not
more fortunate than Johnson; but I found
that the area in front and on one side of
the chapel had been made into a kitchen-
garden; and from what I saw myself, it
is probable that the house of God serves
as the house of the gardener, and that he
there keeps his carrots and turnips during
winter.

By way of compensation, however, I
viewed at my ease, and this was not diffi-

cult, the ruins of the Cathedral and the
adjoining palace [Castle] or residence of the
archbishop. Both these great edifices stood
on raised ground, which looks over the sea.
The palace was, indeed, so close to the water
that the waves have undermined a part of
its foundations.

The Cathedral, to judge from its remains,
without taking into account some adjoining
chapels, and a kind of cloister, as well as
other subordinate buildings around it, was
three hundred and fifteen feet long, and
sixty feet broad. It is one of the most
remarkable and interesting ruins that can
be seen; not only bearing the impression
of time and neglect, but combining also
the most striking evidence of a religious
and fanatical fury which rose to the most
abominable madness and phrenzy.

Towers of the most solid construction
overthrown; columns torn down; huge
gothic windows mutilated, and hanging
as it were in the air; pyramidal steeples,
more than a hundred feet high, and so
solidly built, that as they could not be
knocked down, they were pierced through
and through and indented in every direc-

tion; winding stair-cases that seem to have
no connection with anything; altars over-
turned and heaped one upon another under
the remaining arches; fragments of friezes,
capitals, broken entablatures, mixed up
with sepulchral inscriptions and battered
tombs; the wreck of cloisters, chapels, porticos; columns still standing amid so much
havoc: such in brief is the picture presented by these vast ruins, which fill with
astonishment and chill with horror the
mind of one who looks on them for the
first time.

The traveller is lost in conjecturing
whether it was a terrible earthquake, a
long siege, or an invasion of barbarians,
that brought about so much devastation.
A square tower an hundred feet high,
finely built, and in perfect preservation,
rises intact and alone by the side of these
grand ruins.* It is difficult to account for
this contrast.

In view of such a scene one is soon
led, in spite of oneself, into a train of

* [The tower of the little chapel of St Regulus (Rule),
by much the oldest building of the place, dating as it does
from some interval between the tenth and twelfth centuries.]

melancholy reflections, on the maladies of
the human mind, which degenerate into
madness and mortify the reason.　Are
these frenzies, these deliriums of the under-
standing, like corporeal diseases, inseparable
from humanity? If they are so, man is,
on the whole, the most ferocious, and at
the same time the most unfortunate of
animals; this life might be renounced at
once, were it not for a few chosen individuals who make us endure it.

I was assured that the quadrangular
tower [St Rule's] which rises still entire
amidst so much ruin, has existed for more
than eleven hundred years. It was probably a light-house in former times; at
present it is a memorial only of the feudal
rights which the king has over the city;
and on this account it is preserved with
great care. I mounted by an inside staircase to the highest platform; whence there
is a view of a wide extent of country.

Blaueuw has inserted in his large atlas *
some very accurate engravings of the
principal monuments of St Andrews, at

* [In the *Grand Atlas de Géographie ou Theatrum Mundi*,
1663-67.]

the time when they still existed in all
their splendour. Mr Cleghorn, the professor of history in the University, assured
me, that the materials which had been
furnished to Blaueuw were correct.

These same monuments, in their ruinous
state, have been carefully engraved in four
plates, by Pouncy, from striking drawings,
by J. Oliphant. I saw a collection of them
at the house of the college librarian, who
would not sell them to me. He relig-
iously preserved them under glass, and told
me they were now scarce, and hardly to
be met with for sale.

Before the fanatics, maddened into fury
by the homicidal sermons of the gloomy
Knox, had carried the torch of destruction
to men and things, in this unfortunate town,
it was a place of some eminence; letters and
the sciences flourished within its walls, while
rich and numerous establishments were dedicated to public instruction.

The blows which barbarism dealt on
St Andrews, suddenly changed its appearance. It needs ages to build, but only an
instant to destroy. This city, notwithstanding the length of time which has elapsed

since its misfortunes, appears as if it had
been ravaged by the pestilence. The streets
are large and commodious; but grass grows on
them everywhere. All is sad and silent; the
people living here in ignorance of commerce
and the arts, present a picture of indifference
and languor. This state of inactivity has
its correspondent effect on the state of the
population; for though the place could still
hold fourteen or fifteen thousand souls, it does
not contain at most above three thousand.

I therefore join in the opinion of Johnson,
who, indignant at the neglect in which the
English government leaves establishments dedicated to instruction, exclaims, " It is surely
not without just reproach that a nation, of
which the commerce is hourly extending, and
the wealth increasing, denies any participa-
tion of its prosperity to its literary societies;
and while its merchants or its nobles are
raising palaces, suffers its universities to
moulder into dust."

*Some Objects of Natural History in the
environs of St Andrews*

The cliff on which the Castle of St
Andrews was built, is in many places at least

one hundred feet above the level of the sea;
and the town itself, though standing on a
plain, has the same height above the water.

This huge precipice consists of beds of
white quartzose sandstone, separated at intervals by thin horizontal layers of black
argillaceous shale, soft, a little shining, and
deriving its colour from impalpable particles
of coal.

Where the sandstone comes in contact
with the shale, the first is always divided
into small easily separable layers, somewhat
coloured with coaly particles. There also
may be distinguished some small bits of
wood converted into coal.

To these alternating beds of sandstone,
coloured with carbonaceous material, and of
black argillaceous shale, succeed thick banks
of white sandstone, interstratified in turn
with thin partings of black shale and coloured sandstone; but here the carbonaceous
particles are more abundant.

In short, under the deepest beds of sandstone where the sea has uncovered them, there
can be seen seams of coal, almost pure and fit
for burning.

Industry is in such a state of stagnation in

this town that no person has attempted, by
following these remarkable indications, to
sink a pit, or even so much as to make a
boring for coal which here presents itself
with such promising appearances, and which
from its situation on the edge of the
sea, would form a source of riches to the
country.

I expressed my astonishment on the
subject to several intelligent persons, who
excused this negligence, by saying, that
three or four miles inland some coal-pits
were now worked, and were sufficient for
the supply of the country.*

The sea, notwithstanding the barrier opposed to it by the cliff-bound plateau of
sandstone on which St Andrews is built, has
gained upon the land so perceptibly, that, as
I was assured, there is evidence that within
less than two hundred and fifty years, it has
battered and undermined the rock with such
activity as to destroy almost the whole of the
site of the ancient archiepiscopal castle. A
road that led from the castle to a pier still

* [Those who desire more recent information on the coal-
bearing rocks of this region will find it in the Geological
Survey Memoir on " The Geology of Eastern Fife," 1902.]

existing has been carried away, so that the water completely intercepts the passage; and it should be remarked, that the space destroyed between the castle and the point of the pier is about five hundred toises. Thus have the waves in so short a period wasted away a very considerable area and thickness of solid rock; and at low-water nothing is to be seen but rubbish and ruins on the floor of this restless sea.*

From this encroachment of the waters, however, we are not to draw general conclusions respecting the advance or retreat of the ocean. Purely local conditions have here determined this accidental invasion, which I regard as completely unconnected with any general theory.

I made a careful examination of the place, and recognised some of the causes which have contributed to so much degradation.

And first, the facility which there has always been of procuring stone here, and of removing large blocks of these sandstones, when the sea at ebb-tide lays bare the beach in front of this shore-cliff, is one cause,

* [The waste of this piece of the coast-line is noticed in the Editor's " Scenery of Scotland," 3d Edit. p. 60.]

which is not to be rejected, if we reflect
that the immense quantity of materials employed in constructing the cathedral, several
large churches, convents, the castle and the
houses of the town (formerly much more
numerous), has been taken from this escarpment. I myself saw a large number of
workmen employed in cutting out pretty
large blocks of stone for some repairs they
were making on the pier.

On the other hand, the position of the
beds, the various substances of which they
are composed, and their unequal hardness,
tend to accelerate their degradation. The
coast is so steep that the deep excavation
which extends from the castle to the end of
the pier, bears the name of *Precipice*.

The beds of sandstone lying on seams of
argillaceous shale, which is soft, pyritous and
apt to be weakened by water, are liable to
slide and lose their balance. The upper
beds in falling disturb the others. This per-
manent cause of destruction, combined with
the action of frost, of the atmosphere, and of
alternations of wetness and dryness, must
at length give rise to extensive waste. But,
what is very remarkable and worthy of

attention, all this rubbish is caught up by the sea and exposed to the powerful action of waves and tidal currents, so that the stones dashed against each other or rolled upon the hard and rugged bottom, are soon reduced to powder. Hence considerable deposits of sand are cast up by the sea along its shores, there to form dunes with the help of the winds. Thus the waves which tear up the sandstone and remove it from the coast in large hard blocks, throw it back on a neighbouring shore in the form of sand, which may in time become fertile soil.

It is easy to recognise the identity of this sand, which is intermixed with some coal and argillaceous matter, with the sandstone, which gave rise to it. Such a newly-formed sandy beach covers here a space of four miles in length, and half a mile in width.* Such is probably the origin of most sands, which may in course of time, and in certain favourable circumstances, be a second time formed into sandstone.

I ought to state, before leaving the es-

---

* In these sands are found various living shells. The large razor-fish or *Solen*, and some forms of *Cardium* are common.

carpment of the " Precipice," that the lower
strata, which support a mass more than
eighty feet thick of sandstone and shale,
are themselves remarkable, in that their
beds of sandstone are very hard, and contain
pebbles of different forms and sizes, with a
reddish crust.   When these pebbles are
broken, they are easily seen to have been
derived from a black basaltic and still
magnetic lava, their crust having undergone
alteration.*

   As the rounded lavas thus imbedded are
seen in great number in these lower beds of
sandstone, and as it is to be presumed that
those strata which have been swept away
by the sea contained similar pebbles, it is
beyond doubt that these pebbles of lava
existed prior to the formation of the sandstone; that is to say, that volcanos furnished
these lavas, and that the sea had rounded
them before the sandy materials became
united and consolidated into a mass of sandstone.

   There is no room for any doubt or hesita-
tion respecting the nature of the materials.

   * [Most, if not all of these pebbles were probably nodules of
clay-ironstone which are abundant in some of these strata.]

These pebbles of basalt are so many aids to discovery, and useful indications to those who endeavour to peruse the great book of nature. But this is not a fit place to expatiate further upon the subject. I shall only say, that if accidental circumstances of this kind cannot determine approximately the time which has elapsed since the formation of these lavas and of the sandstone, in which they are inclosed, at least we cannot but believe that both the one and the other must date back to a prodigiously remote epoch.

CHAPTER XIII

*Departure from St Andrews.—Largo.—
Leven. — Dysart. — Kirkcaldy. — Kinghorn.—Leith.—Return to Edinburgh*

SCARCELY had we left St Andrews and
entered on the road to Largo, when we
found the fields scattered over with very
large blocks of basalt. The farmers have
inclosed their lands with them, and thus
afforded to naturalists an easy opportunity
of examining them.*

They are of a fine black colour, great
hardness, and a pure and homogeneous substance. I attentively examined a great number of these blocks which had but recently

---

* [The "erratic blocks" were probably still more numerous
here in the author's day than they are now, the progress of
agriculture having led to the removal of many of them, though
a plentiful crop may still generally be seen. They are mostly
derived from the dark heavy basalts and diabases of Fife,
mingled with pieces of andesite and conglomerate from the
Ochil Hills, and blocks of the crystalline rocks from the
Highlands. This subject is more fully discussed in the
Geological Survey Memoir on " The Geology of Eastern Fife "
already cited.]

been broken, to try whether I could find
any extraneous body in their constituent base.
But I found their substance in general very
pure, and only met with a single fragment
which contained some crystals of black schorl.
The schorls are, in general, very rare in
the volcanic products of Scotland and the
Hebrides.*

After a ride of three miles, we reached a
pretty high plateau, entirely covered with
blocks of basalt, which seem to have been
scattered about at random, and which obstruct
cultivation, for it would not be an easy
matter to remove them.   This high plain is
extensive; oats and rye are gathered on it;
though the vegetable mould cannot be much
more than five or six inches in depth.

This cultivated soil reposes on blackish
argillaceous shale, disposed in strata.   Beds
of sandstone, like those of St Andrews,
succeed the shale, and then follow at a considerable depth seams of excellent coal.   The
number of open pits along the road, shows
that the collieries are worked with great
activity.   I counted more than fifteen within
the space of a mile.

* [See note, vol. 1. p. 208.]

Largo is only a small village; we stopped there to bait our horses. Beds of sandstone of great thickness are exposed to view on all sides; they are overlain with enormous blocks of basalt. I had not before seen in the volcanic parts of Scotland, detached masses of basalt of so great a bulk. This compact lava is very pure and sound, so that it might be formed into large tables or even statues.

Leven and Dysart are pretty large villages, which lie on the road by the sea-side. Coal is worked there, which employs a great many hands. The collieries are carried on upon a greater scale than those in the neighbourhood of St Andrews, and conducted with greater intelligence and more extensive means. Those of the inhabitants who are not employed in the collieries apply themselves to fishing, in which they are very skilful.

Kirkcaldy is a considerable burgh. All the country round it is strewn with blocks of basalt; and this lava-train stretches from Largo to beyond Kirkcaldy, along a space of more than twenty-four miles in length, and eight miles in breadth. What terrible

revolution was it that could seize these
boulders, roll them till they became rounded
and then scatter them over so large a
surface?

I had already seen in the Vivarais, a state
of things in every respect similar; but upon
a plain much higher, above the hill of Maïre.
The masses of basalt are there equally large
and not less numerous. They may be traced
to the little town of Pradelle, through an
extent of more than twenty miles long, and
four or five miles broad. Such a resemblance ought not to be passed over.

From Kirkcaldy we pursued our road to
Kinghorn, a burgh on the water's edge.
The blocks of basalt seemed to multiply as
we approached this place. But here the
basalt is found actually in place, that is,
arranged in great currents such as were
poured forth by the volcanoes.

Between Kirkcaldy and Kinghorn, at a
little distance from the road, three upright
rude stones, have been erected as a memorial
of some event, of which all trace is now
lost. They consist of a coarse-grained
yellowish sandstone. The highest rises a
little more than fifteen feet above ground,

and must have at least five feet of its length
below the soil; the other two are not so
large. (Plate VII.). They appear to be of
high antiquity. Are they the work of the
Romans? I doubt it; for that warlike
people, at the time they over-ran England
and attempted to subjugate the Caledonians,
who gave them the most vigorous resistance,
were too familiar with the arts, to raise such
rustic monuments which have no inscription, nor any mark of workmanship. It is
possible that these rude columns may have
belonged to the religion of the ancient
Druids, or they may have been erected by a
warlike people little skilled in the arts, in
remembrance of some event, or of some great
action the story of which has not come
down to us.

Monuments of this kind are numerous in
Scotland and among the Hebrides. The
natives have various and doubtful opinions
upon the subject. Some call them altars,
temples, or monuments of the Druids; while
others, regarding them as more ancient,
believe that they were erected in the time of
Fingal; that is, at an epoch indeterminate,
and perhaps fabulous; while yet another

Ancient Monuments upon the Shore, between Kirkaldy and Kinghorn.

view is that they are Roman tombs, containing the ashes of illustrious warriors, who fell
in combats with the Caledonians. I shall
leave the unravelling of this enigma to the
Society of Antiquaries of Edinburgh, merely
recalling the fact that similar monuments are
found in Lower Brittany, and that the language of the Bas-Bretons has a strong
affinity to that of the Hebridians. So let us
wait until the Society gives us some explanation upon a subject of antiquity so worthy
of being thoroughly studied.

There are twenty-seven miles from St
Andrews to Kinghorn. It is necessary to
use the same horses for the whole way;
because there is no intermediate place where
they can be changed. Kinghorn is situated
on the edge of the sea; and is the place
of embarkation for the ferry across the
Firth of Forth to Leith, which is close to
Edinburgh.

The beach of Kinghorn, as well as that of
the whole of this coast, is bordered with
currents of lava; some in the form of basalt,
massive and in prisms; others as gravelly
and decomposed lava. These several streams
of volcanic matter rest directly on an argil-

laceous shale which most frequently overlies seams of coal.

In the lavas of Kinghorn I found some zeolite, and a great deal of calcareous spar, adhering to decomposed lavas.*

The length of the passage from Kinghorn to the port of Leith is seven miles. We performed it in two hours, on a tolerably comfortable ferry-boat, which leaves regularly at certain hours. In the middle of the Firth a very rapid current is always perceptible, the sea even in the calmest weather is always agitated there.

The harbour of Leith, when we entered it, was full of vessels, English, Scottish, American, &c. I saw several vessels be-

---

* [It is much to be regretted that Faujas did not halt a little on this coast. He quickly recognised, indeed, the true volcanic nature of the massive and columnar lavas, and he was the first geologist who ever did so. But had he remained to study that wonderful display of ancient volcanic activity, with its included memorials of marine and lagoon life, both plant and animal, he might have forestalled the geologists of the next century by furnishing evidence that might perhaps have stemmed the torrent of Wernerianism, which soon afterwards invaded Scotland. The very rocks on this shore, of which he at once detected the volcanic origin, were afterwards appealed to as evidence for the doctrine of the aqueous origin of basalt. The structure of this most instructive coast-line is given in detail in the Geological Survey Memoir on "The Geology of Central and Western Fife." Chaps. vi. and vii. 1900.]

longing to Glasgow and Leith, which were coated over with bitumen or tar, extracted from pit-coal at the manufactories of Lord Dundonald, who has introduced the making and using of this tar on a great scale in England. The vessels covered with it appeared of a fine shining black, which distinguished them from the others. Several ship-masters from the West Indies whom I questioned, assured me that their vessels thus tarred, arrived in the best possible condition, and were free from worm-holes. Navigation is doubtless much indebted to Lord Dundonald, who has continued with the greatest perseverance to perfect this useful product of coal, and has done everything to bring it into general use in the country—no easy task, when it involves the change of old habits.

We reached the harbour of Leith about six in the afternoon of the 16th of October. William Thornton, who proceeded directly forward from Perth, without accompanying us to St Andrews, waited for us at Edinburgh. We went from Leith to Edinburgh in less than half an hour, along a splendid road.

Thornton had procured us comfortable lodgings at a private house, and at a reasonable rate; for we had determined not to go to Dun's Hotel, where we had been so cruelly fleeced at our first stay in that city. Our new lodgings, for our three selves and our three domestics, cost only eighty-four livres a week.

As we proposed to spend a fortnight in Edinburgh, we made an arrangement with a tavern-keeper who served us with meals in the French style, adding a few Scotch dishes, which we liked. This man was a native of Bourdeaux, and had been brought from France by a Scottish lord, with whom he had been in service for a long time. He afterwards married and settled in Edinburgh. He is a very worthy man, most attentive and obliging. I would recommend his house to naturalists and others who may go to Edinburgh. They have only to ask for the French cook; he is known by that name.

At this table, we made the acquaintance of Baron Hartefeld, whose ordinary residence is at Berlin. He is an estimable man with much intelligence, and travels too for the purpose of gaining knowledge. He had pushed his journeys as far as the Hebrides, and had wished to visit the isle of Staffa; he told us that he ran into the greatest danger in the crossing.

CHAPTER XIV

Edinburgh. — The University. — Learned Societies.—College of Physicians.—College of Surgeons. — Cabinets of Natural History.—Robertson.—Smith.—Black.— Cullen, &c.

EDINBURGH is situated in 55° 57′ of north latitude, and 3° 14′ of west longitude, from the meridian of Greenwich. The distance of this city from London is, by the east road, through Berwick, 388 miles; by the middle road, through Wooler, 378 miles; and by the west road, through Carlisle, 396 miles.

The sciences, literature, natural history and the arts, being essential features in the plan of my journey; what I have to say of Edinburgh will turn chiefly on these subjects: topographical descriptions of this town may be found elsewhere.

The University

The following is the establishment of the University of Edinburgh, with the names of those who were at this time its professors.

The King, is the Protector.
Doctor Robertson,* Principal.
Andrew Hunter,† professor of Divinity.
Robert Cumming, Church History.
Doctor James Robertson, Hebrew.
Andrew Dalziel,‡ Greek.
Dugald Stewart,§ Mathematics.
Adam Ferguson,‖ Moral Philosophy.

* [William Robertson (1721-1793), the historian of Scotland and Charles V., was Moderator of the General Assembly of the Church of Scotland (1763) appointed Historiographer for Scotland; became Principal of the University of Edinburgh in 1762, and retained the office for thirty years.]

† [Andrew Hunter (1743-1809), Professor of Divinity from 1779 till his death; Moderator of the General Assembly, 1792.]

‡ [Andrew Dalziel (1742-1806) became Professor of Greek in the University in 1779, and held the office till his death; compiled two series of extracts from Greek classics; wrote a History of the University.]

§ [Dugald Stewart (1753-1828), a prominent Scottish philosopher, associated with his father in the Chair of Mathematics, 1775; famous for the distinguished politicians who were his pupils; became Professor of Moral Philosophy in 1785, but ceased to give his lectures after 1809.]

‖ [Adam Ferguson (1723-1816), first an Army Chaplain, and present with Black Watch at battle of Fontenoy; Pro-

John Robison, Natural Philosophy.

Alexander Fraser Tytler,* Civil History.

William Wallace, Scots Law.

Robert Dick, Civil Law.

Allan Maconochie,† the Law of Nature and Nations.

Hugh Blair,‡ Rhetoric.

John Hope,§ Medicine and Botany.

Francis Home,‖ Materia Medica.

William Cullen,¶ Practice of Medicine.

fessor of Natural Philosophy at the University, 1759; Professor of " Pneumatics and Moral Philosophy," 1764-85.]

* [Alexander Fraser Tytler (1747-1813) published various historical works; Professor of History, 1780; became a judge of the Court of Session in 1802 with the title of Lord Woodhouselee.]

† [Allan Maconochie (1748-1816) became Professor of Civil Law in 1779, and held the office until he was appointed to a judgeship in the Court of Session, when he took the title of Lord Meadowbank. He wrote some legal and agricultural works.]

‡ [Hugh Blair (1718-1800), a noted divine in the Scottish Church; was the first Professor of Rhetoric and Belles-lettres in the Edinburgh University, 1762; espoused the genuineness of Macpherson's " Ossian " and became its chief defender.]

§ [John Hope (1725-1786), appointed Professor of Medicine and Botany, (1761-1786); edited the *Genera Animalium* of Linnaeus; founded the Edinburgh Botanic Garden.]

‖ [Francis Home (1719-1813) was Surgeon of Dragoons in the Seven Years War; appointed to the newly founded Chair of Materia Medica in 1768 which he filled till his death.]

¶ [William Cullen (1710-1790), further noticed, *postea*, p. 242, was at the head of the medical profession in Scotland. He became Professor of Chemistry at Edinburgh in 1755, and exchanged that Chair in 1766 for that of Medicine or the

James Gregory,* Institutes of Medicine.
Joseph Black,† Medicine and Chemistry.
Alexander Monro,‡ Anatomy.
Alexander Hamilton,§ Midwifery.
John Walker,‖ Natural History.

Royal Society

The Duke of Buccleugh is President of the Royal Society.

Theory of Physic. He was elected Fellow of the Royal Society in 1777.]

* [James Gregory (1753-1821), educated at Aberdeen, Edinburgh and Oxford; became Professor of the Institutes of Medicine at Edinburgh University, 1776; and succeeded Cullen in the Chair of Medicine in 1790.]

† [Joseph Black (1728-1799), one of the most illustrious chemists of his time, and one of the fathers of modern chemistry; was Professor of Medicine at Glasgow, 1756-66; carried on many physical experiments and greatly contributed to the success of Watt's labours on the steam-engine; succeeded Cullen in the Chair of Chemistry and Medicine at Edinburgh, 1766-97. See *postea*, p. 235.]

‡ [Alexander Monro (1733-1817), usually known as *secundus* to distinguish him from his father (*primus*) and his son (*tertius*) who all successively held the Chair of Anatomy in the University of Edinburgh. He succeeded his father in 1754, and was followed by his son in 1798. He published some important anatomical observations.]

§ [Alexander Hamilton (1739-1802) held the Chair from 1780 till 1800.]

‖ [John Walker (1731-1803), best known as a botanist; appointed Professor of Natural History in 1779, and held the Chair until his death, when he was succeeded by Robert Jameson, the champion of Wernerianism in Britain.]

Henry Dundas,
Thomas Miller (Lord- } Vice-Presidents.
Justice-Clerk),

[The Councillors are]

Mr Baron Gordon.

Lord Elliock.

Major-general Fletcher Campbell.

Mr Commissioner Adam Smith.*

Mr John Maclaurin.

Doctor Adam Ferguson.

Doctor Alexander Monro.

Doctor John Hope.

Doctor Joseph Black.

Doctor James Hutton.†

Professor Dugald Stewart.

---

* [Adam Smith (1823-1790). This great political econo-
mist, author of the "Wealth of Nations," became Professor of
Logic in the University of Glasgow, 1751 ; and next year was
transferred to the Chair of Moral Philosophy there, but eventually gave up his connection with that University, travelled on
the Continent as tutor to the young Duke of Buccleugh from
whom he afterwards received a pension ; lived in Edinburgh
in his later years, and was one of the notable men in that city
during the later decades of the eighteenth century.]

† [James Hutton (1726-1797), one of the most illustrious
founders of Modern Geology ; after some years spent in agricultural pursuits on his property in Berwickshire, he settled in
Edinburgh and devoted himself to philosophical and scientific
enquiries.   At the time of Faujas' visit, Hutton was engaged
on the first draft of his immortal "Theory of the Earth."
See p. 235.]

Professor John Playfair.*
Professor John Robison, Secretary.
Mr Alexander Keith, Treasurer.

*Society of Antiquaries.— College of Physicians.
—College of Surgeons.—Medical Society*

A society has lately been established in this city for the purpose of collecting and preserving every thing that relates to Scottish antiquities. The Earl of Bute is the President, the Earl of Buchan is First Vice-President, and Lord Gardenstone is Second Vice-President.

There are besides a College of Physicians, a College of Surgeons, and a Medical Society.

What is called the High School of the city, is a popular institution which shows that here nothing connected with public instruction is neglected. Several of the masters are employed in elementary teaching.

* [John Playfair (1748-1819) entered the Church of Scotland; but having a strong mathematical bent, ultimately gave up his clerical duties to become joint-professor of Mathematics in the University of Edinburgh (1785); in 1805 he was transferred to the Chair of Natural Philosophy which he held to the end of his life. He wrote a treatise on Geometry which went through many editions, but he is best known for his " Illustrations of the Huttonian Theory " in which he developed with admirable perspicuity the leading doctrines of his friend and master Hutton, and thus greatly contributed to the advancement of Geological Science.]

From all these institutions it can be seen that literature, science and art are here held in high esteem. The city is honoured, also, by the great men it has produced in almost every kind of distinction; while the fame of the professors has attracted to its walls foreigners from all parts of the world, and has given lustre to the town as well as an increase in the ease of living.

Edinburgh, from its situation and the calm which pervades it, is a proper place for the sciences; these love neither tumult, nor parliamentary discussion, nor the noisy bustle of commerce, nor such multiplied objects of distraction and pleasure as are found in London. From time immemorial the muses have fixed their home on a hill, beside a solitary fountain.

Mentioning the muses, brings to my recollection an inscription in their honour, which is placed above one of the gates of the University. It is rather extraordinary:

MUSIS ET CHRISTO

TO THE MUSES AND CHRIST

This association may appear a little profane, but it is only a *tour de force* on the part of the author of the inscription by which he probably meant to announce, in lapidary style, that both letters and religion are taught here. A presbyterian minister, who accompanied me on my visit to the University, wished to justify this singular inscription, which seemed to him very happy and very striking: he asked what I thought of it.

I replied with a smile, that I believed the inscription might be interpreted in a more serviceable way, if the meaning which I was inclined to give it, were adopted. The muses, in presiding over an establishment which raises man to his true dignity by instructing him, are there, said I, to beseech reason to proscribe two chairs, the one called *Theology* and the other *Church History*; and out of those of *Logic, Moral Philosophy, Natural Philosophy, the Laws of Nature and Nations, Civil Law*, and *Scots Law*, to make only one single chair, that of the *Laws of Nature and Nations*.

On the other hand, the greatest of moralists, is placed beside the muses, to remind the inhabitants of these countries that true knowledge is the enemy of fanaticism and intolerance; that those who have caused so much of the blood of Scotland to flow on account of religious opinions were unin- fluenced by any morality and had no religion, and that those who overthrew and destroyed, from top to bottom, so many ancient monuments, because they had been dedicated to a religion for which the people no longer cared, were real barbarians, as ferocious as they were ignorant.*

Cabinet of Natural History

The cabinet of natural history, in the University, is under the direction of Doctor Howard. This collection gave me more pleasure, and interested me more than that of the British Museum in London, though it was far less considerable; but, the objects which compose it, are in more methodical order, particularly the stones and minerals. They have here had the good sense to collect all the productions of Scotland which they have been able to procure.

Thus this museum is as instructive and interesting to the inhabitants of Scotland, as

* Such as a Knox, who entitled an account of the assassina- tion of Cardinal Beaton, "The Joyous Narration," &c. ; such as a Beaton himself, who burnt human beings alive, then called heretics.

it is attractive to foreigners, who would much rather find in such collections a display of the natural and local riches of a country, than most of the disconnected and inconsequential objects brought from India, which one sees repeated over and over again in almost every cabinet.

The place which contains the natural history collection of the University ought to be larger and decorated with a little more taste, and the systematic classification should also be as well done in the other parts as it has been in the minerals. These improvements will certainly one day come; if they have not been made already, we must remember that the old universities made no provision for the natural sciences. It is only a short while ago that a chair for this subject was established here.* We must hope that as a taste for this delightful study increases in a city which the sciences have so long made their home, an effort will be made to provide the Natural History Museum of Edinburgh with a building more worthy

* [The first professor of Natural History in the University of Edinburgh (Robert Ramsay) was appointed in 1770. He was succeeded by John Walker who has been already referred to (p. 227).]

of a country which can supply the greatest wealth of material for the illustration of this subject. I therefore invite Doctor Howard, who possesses much knowledge and zeal, and who loves his country, to solicit from the government the grant of a suitable building, with grounds sufficiently extensive to allow of the natural history collections and the botanic garden being placed together.*

Lithology, and the study of minerals, have not as yet made all the progress which they will one day attain in Scotland. There are therefore few collections of this kind. Doctor James Hutton is, perhaps, the only person in Edinburgh who has brought together in a cabinet some minerals and a large series of agates chiefly found in Scotland; but I observed, that he had not been sufficiently careful to collect the different matrices in which they are enclosed, and which serve as a complement to the

* [The more ample building which the author hoped for has been provided in the " Royal Scottish Museum," where the University collections form part of an extensive series of natural history specimens. The mineral, petrographical and palæontological cases are of especial interest and value, and care has been taken to give the productions of Scotland in these branches a foremost place.]

natural history of these stones. I therefore
had much more pleasure in conversing with
this modest philosopher than in examining
his collection, which presented me with
nothing new, since I had lately seen and
studied in place the greater part of the
objects in this collection.

Doctor Hutton was at this time engaged
in the calm of his study, writing a work on
the theory of the earth.*

During my residence in Edinburgh I
visited, as often as possible, the celebrated
chemist, Dr Black, who in 1761 gave the
first exact analyses of calcareous earth, in
which he demonstrated the existence of the
gaseous acid, popularly known as "fixed
air" [carbonic acid]. This illustrious philo-

---

* This work, which is rather a memoir containing general
views of the subject than a body of observations on the theory
of the earth, appeared in 1785, in the *Transactions of the
Royal Society of Edinburgh* for that year, under the following
title: "Theory of the Earth, or an Investigation of the Laws
observable in the composition, dissolution, and restoration of
land upon the globe, by James Hutton, MD, FRSE, and
Member of the Royal Academy of Agriculture at Paris."
[This classic of geological literature was afterwards expanded
by its author into two volumes which were published in 1795.
A third volume was left in MS. but was not published; the
manuscript of it appears to have been lost, except six chapters
which came into the possession of the Geological Society of
London, and were published by it in 1899.]

sopher honoured me with the most polite and kind attention.

After dining with him one day, he shewed me two pieces of petrified, or more properly, *quartzified* wood ; for, upon examining them with a microscope, it appeared that the quartzose juice had penetrated through all their parts, and given them such a degree of hardness that they struck fire with steel. This wood had been sent to him from Ireland. Its colour was brown, and nearly the same as that of the wood of Mahalep, when it is worked.

This wood penetrated by quartz, possesses the following singular property: If small chips, broken off with a hammer, are thrown on burning coal, in about a minute a pleasant smell is perceived resembling that given off by the wood of aloes.

It is doubtless astonishing that the essential odoriferous oil of this wood should have been preserved during the long time necessary for transforming the wood into the condition of quartzose petrifaction. Even supposing that some particular circumstances may have accelerated the petrification, it is still very extraordinary that this wood, which bears all the marks of a vegetable foreign to these countries, should be found on the banks of Lough Neagh in Ireland.

Doctor Black was pleased to give me the only two specimens he had of this wood, telling me that he did not collect objects of this kind, and that he would be charmed that they should find a place in my cabinet.

This savant also shewed me the mechanism of a portable furnace of his own invention, which will prove of great utility in the arts, and in chemistry. It is so contrived that the heat may not only be regulated at pleasure, but carried to such a high degree as to melt iron-nails. The theory of it is as simple as it is ingenious. The invention may perhaps even be applicable to high furnaces in which iron-ore is smelted.

It is in the manner in which the thing is lined, that the merit of the invention chiefly consists. The furnace is made of thick iron plates, and differs very little in its structure from the ordinary stove. Its form is cylindrical, and a cover is fitted to the top, which is occasionally taken off for the supply of fuel.

The air enters through holes of different

sizes formed in a ring which turns round,
and thereby makes it easy to regulate the
draught.   But it is not, I repeat, this construction which constitutes the great merit
of the invention; for I have seen, both at
Paris and in Germany, many furnaces, akin
to this one in their mechanism, as far as
concerns the mode of graduating the admission of air, and of raising the heat to the
degree required.

It is the internal lining and the materials
used in making it, that do honour to the pro-
found knowledge of Doctor Black.

Wood-charcoal is reduced to a fine powder,
and passed through a sieve; some soft clay
is also reduced to powder: the colour of the
latter is of no consequence; the least fusible
and the most refractible kind is best.

The clay is mixed with water in a trough,
in the measured proportion of a fourth part
of clay to three-fourths of charcoal dust.
The whole is then well kneaded and amalgamated, and the paste is left somewhat liquid.
If the clay is very adhesive, the proportion
of charcoal can be increased.   The mixture
being thus prepared, a certain quantity of it
is spread over the inside of the furnace as a

thick coating, the material being repeatedly applied and smoothed with the hand so as to make it as equal as possible. This first plaster is made about a line thick, and allowed to dry slowly, without fire, so as to avoid the risk of cracking. When it has become hard, and the hand can be passed over it without removing anything, a second layer is added, which being allowed to dry, a third is laid over it in the same manner; and so on in succession, until the coating is about an inch thick.

Great attention must especially be paid to drying the layers slowly and forming them into one body, to which the fire will afterwards give great consistence.

It is well known to physicists and chemists, that charcoal is one of the worst conductors of heat. Founders, blacksmiths, and other workmen, have learnt by long practice, transmitted from father to son, to make use of charcoal dust in many of their operations, without thinking about the manner in which it acts. The useful effects of this dust arise less from its combustible nature than from its being a bad conductor of heat; or rather from its

capacity of retaining heat, concentrating it,
and preventing its escape and loss from the
surrounding sides.

I have thought that the details I have
here given may prove useful to the arts,
and that those who love and cultivate
this branch of applied science, may make
successful applications of the invention.
These motives should excuse the length of
this article.

I had several times the opportunity of
seeing Dr John Aiken, a private teacher of
Anatomy, in Edinburgh. He shewed me
a number of ingenious machines of his
invention; among others, one for facili-
tating difficult births, which had nothing
in its use either terrifying or dangerous, for
the inventor had followed nature as closely
as possible.

This instrument may be compared to a
long slender hand. It is introduced quite
open and without any kind of compression
into the womb of the mother. This artificial hand, which is covered with a fine
soft skin, is placed against the child, upon
which it is made to collapse to the degree
of contraction wanted, by the means of a

screw in the handle, which acts with a gradual and gentle motion. The accoucheur then using his right hand, aided by this point of support, may deliver a woman in difficult labour with much more facility than in any other manner. Doctor Aiken assured me, that he had experienced the greatest success in the use of this instrument.

The knowledge of whatever may contribute to relieve suffering humanity, ought to be as widely diffused as possible. I therefore begged Doctor Aiken to allow me to have a similar instrument which I might take to France as a model; and he had the complaisance to get his best workman, who in a few days completed a perfect duplicate which I proposed to submit in Paris, to the examination of some eminent practitioners in midwifery.

Doctor Aiken also shewed me a gun with a single barrel which discharges two shots; but while I admired his inventive genius, I could not avoid telling him, that I was far better pleased to see so skilful a surgeon employed in healing than in destroying.

Some days after, I had the pleasure of

dining with Doctor Cullen, who, perhaps,
is the oldest, and certainly is one of the
most celebrated physicians of Europe. The
science of medicine owes him great obligations, and the city of Edinburgh will not
forget that the reputation of Cullen has
attracted within its walls a multitude of
foreigners, who have come from all parts
of the world to receive instruction in that
learned school.

Doctor Cullen lived in the midst of a
numerous family, forming around him a
circle of lively and loveable friends. His
house was pervaded with good nature and
urbanity. He well deserved all these advantages, for his own manners and disposition were of the most agreeable kind.
I found that in his behaviour and mode
of living, he had some resemblance to
Buffon, which made him doubly interesting
to me. His table was plentifully served, but
without any luxury. I was, however, somewhat astonished, after the dessert and before
the tea and coffee, to see them bring in a
profusion of punch.

This regimen, in the house of a physician
of such a reputation, appeared to me a

little strange. He noticed this, and said
to me with a smile, that this drink was
not only suited to his age, but that a long
experience had convinced him that, taken
in moderation, it was very salutary for
the inhabitants of Scotland, particularly
towards the end of autumn and in winter,
when the cold damp, which then generally
prevails in this climate, prevents the equi-
librium of the perspiration. " Punch," he
told me, " is a warm stimulant, which op-
erates wonderfully in maintaining or re-
establishing that necessary secretion."

This humid and penetrating atmosphere
had, for some time, affected myself in a very
disagreeable manner, notwithstanding the
active life which I led. I am persuaded,
that it is one of the causes of that sombre
melancholy which so often affects the English. In vain I took exercise, in vain I tried
to divert myself pleasantly in the enquiries
and occupations suited to my tastes; I
found that the mists, the frequent rains,
the daily winds, passing suddenly from
heat to cold, a certain sharpness in the
air, which one feels better than one can
describe it; above all the disappearance of

the sun, which fogs or clouds constantly
eclipse at this season, plunged me into
an involuntary melancholy, which I should
not have been able to endure for long.

From time to time, to raise my spirits,
I was told that the sun was about to appear; but I was more than once tempted
in my ill humour, to reply to them, as
Caraccioli, the viceroy of Sicily, did to an
English nobleman, who desired him to
look at that luminary in London; "Your
English sun, my lord, very much resembles
our Sicilian moon."

Tired to find myself in this condition,
I at last adopted the regimen of Doctor
Cullen. Each day after dinner I took a
glass of punch, composed of rum, sugar,
lemon juice, a little nutmeg and boiling
water, and I soon found myself quite
well.*

I saw at Edinburgh several other savants
and men of letters, among whom were
Dr Anderson, Sir John Dalrymple, and
the historian, Principal William Robertson,

* This excellent physician is no more. He was regretted
by his friends, and mourned by the city of Edinburgh which
erected a funeral monument to his memory. He was worthy
of that honour, and the city was worthy of him.

with whom I enjoyed various conversations.

That venerable philosopher, Adam Smith, was one of those whom I oftenest saw. He loaded me with polite attentions, and sought to procure for me every information and amusement that could interest me in this town.

Smith had travelled in France, and resided for some time in Paris. His collection of books was numerous and well chosen: all our best French authors occupied a distinguished place in his library. He was very fond of our language.

Though advanced in years, he still possessed a fine figure. The animation of his features was striking, when he spoke of Voltaire, whom he had known and whom he greatly liked. "Reason," said he, one day, as he shewed me a very fine bust of this author, "owes him incalculable obligations; the ridicule and the sarcasms which he plentifully poured out upon fanatics and hypocrites of all sects, have prepared men's minds for the light of truth, to the search for which every intelligent mind ought to aspire. He has done more in this matter than the books

of the gravest philosophers, which everybody does not read; while the writings of
Voltaire are in general made for all and read
by all."

On another occasion he observed to me,
"I cannot pardon the Emperor Joseph II.
who pretended to travel as a philosopher,
for passing near Ferney without having
gone to pay homage to the historian of the
Czar Peter I. I inferred that Joseph could
only be a man below the average."

Another time, when I was taking tea with
him, he spoke to me of Rousseau with a
kind of religious respect, "Voltaire," said he,
"sought to correct the vices and the follies of
mankind by laughing at them, and sometimes
even getting angry with them; Rousseau, by
the attraction of sentiment, and the force of
conviction, drew the reader into the heart
of reason. His *Contrat Social* will in time
avenge him for all the persecutions he
suffered."

He asked me one day, whether I was fond
of music? I answered, that it formed one
of my chief delights, whenever I was so
fortunate as to hear it well executed. "So
much the better," said he; "I shall put you

to a proof which will be very interesting for
me; for I shall take you to hear a kind of
music of which it is impossible you can have
formed any idea, and it will afford me great
pleasure to know the impression it makes
upon you."

Next morning at nine o'clock, Smith came
to my lodgings. At ten he brought me to
a spacious concert-room, plainly but neatly
decorated, and full of people. I saw, however, neither orchestra, musicians, nor instruments. We sat waiting for more than
half an hour. A large empty space in the
middle of the room was surrounded with
benches which were occupied by gentlemen
only; the ladies were dispersed among the
other seats. "These there," said he, allud-
ing to the gentlemen who sat in the middle,
"are the judges of the competition which is
about to take place among the musicians.
Almost all of them are landlords living in
the Isles or Highlands of Scotland; they
are thus the natural judges of the contest;
they will award a prize to him who shall
best perform a piece of music which is a
favourite with the Scots. The same air
will be played by all the competitors,

no matter how many of them there may
be."

A few moments later, a folding door
opened at the bottom of the room, and to
my great surprise, I saw a Scottish High-
lander enter, in his costume of Roman
soldier, playing upon the bagpipe, and walking up and down the empty space with
rapid steps and a military air, blowing the
noisiest and most discordant sounds from an
instrument which lacerates the ear. The
air he played was a kind of sonata, divided
into three parts. Smith begged me to give
it my whole attention, and to tell him
afterwards the impression it made upon me.

But I confess that at first I could distinguish neither air nor design. I only saw
the piper marching always with rapidity,
and with the same warlike countenance.
He made incredible efforts both with his
body and his fingers to bring into play at
once the different pipes of his instrument,
which made an insupportable uproar.

He received nevertheless great applause
from all sides. A second musician followed
alone into the arena, wearing the same
martial look and walking to and fro with

the same haughty air. He seemed to excel
the first competitor; as I judged from the
clapping of hands and cries of *bravo* that
resounded on every side; grave men and
high-bred women shed tears at the third
part of the air.

After having listened to eight pipers in
succession, I began to suspect that the first
part was connected with a warlike march
and military evolutions: the second with a
sanguinary battle, which the musician sought
to depict by the noise and rapidity of his
playing and by his loud cries. He seemed
then to be convulsed; his pantomimical
gestures resembled those of a man engaged
in combat; his arms, his hands, his head,
his legs, were all in motion; the sounds of
his instrument were all called forth and
confounded together at the same moment.
This fine disorder seemed keenly to interest
every one.

The piper then passed, without transi-
tion, to a kind of andante; his convulsions
suddenly ceased: he became sad and overwhelmed in sorrow; the sounds of his instrument were plaintive, languishing, as if
lamenting the slain who were being carried

off from the field of battle. This was the
part which drew tears from the eyes of the
beautiful Scottish ladies. But the whole
was so uncouth and extraordinary; the impression which this wild music made upon
me contrasted so strongly with that which
it made upon the inhabitants of the country,
that I am convinced we should look upon
this strange composition not as essentially
belonging to music but to history.* It
should be remarked that we find no trace
of a written language among these people,
neither in monuments nor in manuscripts;
whence I presume that they consigned the
memory of the events which interested them
most to this kind of chant, which could be
easily handed down from generation to
generation. Accustomed to hear these airs

* Johnson makes the following observation on an air
which he heard at the seat of Sir Alexander M'Donald in the
isle of Skye: "As we sat at Sir Alexander's table, we were
entertained, according to the ancient usage of the North, with
the melody of the bagpipe. Every thing in those countries
has its history. As the bagpiper was playing, an elderly
gentleman informed us, that in some remote time, the
M'Donalds of Glengary having been injured, or offended by
the inhabitants of Culloden, and resolving to have justice or
vengeance, came to Culloden on a Sunday, where finding their
enemies at worship, they shut them up in the church, which
they set on fire; and this, said he, is the tune which the piper
played while they were burning."

from their infancy, and taught by their parents to connect them with the deeds they commemorate, the Highlanders have imperishable associations with this music, which thus becomes in a manner sacred in their eyes. There need be no surprise therefore that they should be so passionately fond of it. They have, however, another sort of music, better adapted for singing and more in accordance with the rules of art, which they use in their dances and songs; but it is regarded by them as much inferior to the first kind.

The same air was played by each competitor, of whom there was a considerable number. The most perfect equality was maintained among them; the son of the laird stood on the same footing with the simple shepherd, often belonging to the same clan, bearing the same name, and having the same garb. No preference was shown here save to talent, as I could judge from the hearty plaudits given to some who seemed to excel in that art. I confess that it was impossible for me to admire any of them. I thought them all of equal proficiency; that is to say, the one was as bad as the other; and the air that was played as well as the instrument itself, involuntarily put me in mind of a bear's dance.

The competition was followed by a lively and animated dance, formed by one part of the pipers, while the others played suitable airs, which had some melody and character; but the union of all these bagpipes produced an unbearable noise.

The competitors afterwards formed themselves into a line two deep, and marched in that order through a part of the town to the foot of the castle of Edinburgh, which is perched upon a volcanic rock. There they played an air, a kind of ballad, in honour of the unfortunate Mary Stuart for whom the Highlanders, as well as the inhabitants of the Isles, have retained an attachment and a kind of religious respect, which the misfortunes of that queen have only served to increase. They are affected every time they speak of her; they look on her as innocent and as the victim of the cruel jealousy of the implacable Elizabeth. Mary was their Queen. They know that she was beautiful, gentle. affable and generous; that she loved the arts; that she languished in a long and touching captivity; and that she died with resignation and courage. Much less than this is enough to interest honest peaceable men, whom state policy, and the crimes which it engenders, have not yet corrupted, and who abhor the shedding of blood in any way but for legitimate defence.

While the musicians were at the Castle, the judges were engaged in discussing the merits of the several competitors, in order to award the prize to the most worthy: a bagpipe with ivory mountings, a fine instrument with everything complete, or other suitable objects are each year presented to the winner.

I do not know to what period, probably a very ancient one, the institution of these prizes goes back. It is not known if the competition has always taken place in the town of Edinburgh, on account of the distance of the Highlands, or if it was Queen Mary who transferred it to her capital.

They told me during my stay in Mull that there had been beyond all time of memory in that island, a college or society of bagpipers, which was not even entirely extinguished after the death of the famous

Rankin, who had the direction of it about thirty years ago. M'Rimmon kept a similar school in the isle of Skye, and some of the principal families of the Hebrides always kept a piper, whose office was hereditary.

While I remained at Edinburgh I made several excursions for the purpose of examining the natural history of the neighbourhood of that city, and I there made a large collection of all the volcanic products, and other interesting mineralogical specimens. Each article duly labelled, was carefully put into a special packing-case, and Dr Swediaur kindly undertook to send it to France together with the other collections I had made among the mountains of Scotland and in the Hebrides.

This valuable package, the fruit of so much trouble and so much pleasure, was lost with the vessel which carried it, on the coast of Dunkirk. The crew had hardly time to escape in a boat; nothing could be saved, and I was deprived in a moment of a large collection which had for me the greater interest, in that it contained several new objects which would have been highly interesting to naturalists.

Fortunately, whenever I had leisure, I wrote in my journal exact notes of the specimens I collected; these have been most useful to me in describing the precise lithology in the neighbourhood of Glasgow, Perth, and in Staffa, the isle of Mull, and other places. Pressed by my numerous engagements at Edinburgh, however, I did not register the labels that accompanied the whole of the specimens I there collected. This is the only omission of the kind I have to charge myself with; but unfortunately it prevents me from giving a description of the various and remarkable productions which abound in the hills that are grouped around that city, and which have almost all been a prey to the action of the subterranean fires.

I should have been the more desirous of giving a detailed account of this series of remarkable specimens, which leave no doubt as to the existence here of ancient volcanos which once devastated this region, since I found the greater part of the learned men of the city prejudiced against this opinion.

The castle, which commands the town, is built upon a hill formed of compact lava of

the nature of basalt. The black colour of
this lava, and the gothic aspect of the castle
which crowns this volcanic peak, form a
striking contrast with the modern white
houses, built with taste in the quarter of the
New Town.

Not far from this and on another eminence formed of greyish lava, stands a kind
of Greek temple adorned with columns;
this monument, erected by public gratitude
to the memory of a philosopher and an
historian, contains the ashes of Hume.

The whole of the elevated chain which
rises behind the town in the quarter where
the hills seem piled up one against the
other, is composed of basaltic lava. This
material, which at one time must have
been liquefied, has in some parts undergone
prismatic contraction due to cooling. If the
structure has not here the fine regularity
of the prismatic columns of Fingal's Cave,
or the Giants' Causeway of Antrim, this
is probably due to the cooling having been
too rapid, or the want of regularity may
arise from causes still unknown to us.

One of the hills of this chain, by its
form and a depression towards its steepest

side, presents some root of resemblance to an arm-chair, with a gigantic seat, or at least it was anciently thought to have such an appearance, whence it was said to be the seat of the Giant Arthur. This eminence, which has nothing remarkable about it but its height and its escarpment, is known in the old chronicles by the Latin name of *Arthuri Sedes*, and in English by that of *Arthur Seat*. It is possible that the designation may come from another source of which the tradition is lost, for a number of volcanic hills have always borne the names of giants or of attributes connected with these allegorical beings.*

Sibbald, in his *Scotia Illustrata*, printed in 1684, gives an account of a barometrical observation made by the mathematician, George Sinclair, on the summit and at the bottom of this mountain which he calls *Sedes Arthuri*; it is clearly the same as that with which we are here concerned.

In examining the large blocks of basalt which are detached from this mountain

* [The more prevalent explanation has connected the name with the ancient British King Arthur, and an able scholar has endeavoured to show that this district was the scene of some of the Arthurian legends.]

and from piles of rubbish about its base, I observed kernels of zeolite even in the centre of the lava, and collected some fine specimens. This zeolite is white, in some parts shading into a greenish hue; it is neither radiated nor crystallized in a regular manner; but has rather the somewhat scaly texture of white marble. It is hard and susceptible of the most brilliant polish, which is not surprising, when we know that it contains a small mixture of quartzose earth, giving it a somewhat chalcedonic appearance; but it is fusible with the blowpipe, bubbles up in melting, and has all the properties of zeolite.*

Behind these volcanized mountains there are beds of quartzose sandstone, which have been strongly acted on by fire, thereby acquiring a reddish colour. Indeed, the

* [From this account it seems tolerably certain that the author is speaking of Salisbury Crag, and of the debris on its slopes and at its feet. The rock of which that Crag is mainly composed is one of the intrusive sills referred to in Chapter IX. of Vol. I. and not a superficially erupted lava. The mineral here somewhat vaguely characterised was probably Prehnite. Faujas does not appear to have examined Arthur Seat proper; at least if he did so, it is difficult to account for the absence of a description of the fine series of columnar and massive lavas in the east half of that hill, and of the striking agglomerates around its summit.]

traces of the subterranean fires are manifest every where around Edinburgh, with the same characters and the same clearness as in the environs of Perth, Glasgow, and Dumbarton, and in the island of Staffa.

To my infinite regret I can only give here general statements: If my fine and large collection had not been lost, I should have been able to make known and to describe a series of volcanic products which would have removed every doubt, and demonstrated that the district of Edinburgh has been the prey of ancient volcanos, since there are still found in it lavas like those of Vesuvius and Etna.*

* [Every geologist must regret the loss of that packing-case off Dunkirk. Had its contents reached Paris they would have been diligently studied, and there can be little doubt that, even in the author's own life-time, they would have been more precisely determined than was possible in the rapid examination which alone could be given to them in his journey. He might then have been able to publish a second edition of his "Voyage," with all the additional and more definite petrographical information which his collection would have supplied.]

CHAPTER XV

Departure from Edinburgh.—Itinerary to Manchester.—Natural History

AFTER taking leave of Dr Black, Dr Cullen, the learned Smith, and the other respected men who had loaded me with kindness during my stay in Edinburgh, I made preparations for my return, and determined to take the Carlisle road to London, which would give me an opportunity of seeing Manchester, Derby, Buxton, Castleton, Birmingham, Oxford, &c.

We were four when we left London for Edinburgh; we were only two when we quitted Edinburgh on our way back to London. I forgot to mention that M. de Meciès, after visiting Fingal's Cave, the principal object of his journey, left us at Mr Maclean's, in the Isle of Mull, and returned to London, to which business recalled him. Our second travelling companion, the interesting and worthy William

Thornton, intending to pass some months with his family in America, remained in Edinburgh, where he had friends and acquaintances, and where he could await the departure of a vessel. We parted from him with pain, but resolved to cultivate his friendship, for his moral qualities as well as his passion for science made him a most estimable man, well worthy of the attachment of those who make his acquaintance.

We set out then from Edinburgh, Count Andreani and I, and took the Carlisle road on the 3rd of October.

A mile and a half from Edinburgh, the lavas and other volcanic substances which surround that city, disappear; they are then succeeded by quartzose sandstones, which in several places overlie rich seams of coal that are actively worked. This sandy zone, which is pretty extensive, disappears in its turn, and we once more enter into a region of volcanic materials from Lasswade to Selkirk, in passing through Middleton, Bankhouse, Stagehall, Crosslee, &c.*

* [This is one of the most curious errors of observation in Faujas' account of his journey. There are practically no volcanic rocks the whole of the way. What he calls "lavas

The aspect of this part of the country is very wild, and displays only sterility. The black, bluish, and reddish brown lavas which are seen by the wayside, are almost all disposed in tables or plates like some slates, but they have all experienced the action of fire. Some of them are of the nature of basalt; others, less hard, exfoliate and become decomposed in the air.

We passed by Ashkirk, Hawick, Allanmouth, Binks, Mosspaulgreen, Redpath, and Langholm. This last place is marked as 69 miles from Edinburgh. Everything is volcanic from Ashkirk to Langholm. The lavas at Hawick form steep hills, and are disposed in thin horizontal beds, or rather leaves, which resemble slates; only their colour is a little paler. I should not be surprised if this schistose rock, which is of the nature of trap rather than that of slate, had been volcanised in place, and if its fissile structure belongs to its original formation. It shows in some places transverse veins of rose-coloured calc-spar, but these are few and

disposed in tables or plates " are in reality sedimentary deposits consisting of grits, greywackes, and shales of Silurian age, forming the Southern Uplands. See also Vol. I., p. 170.]

accidental.   Even an inexperienced eye cannot fail to see that all the mountains and hills
along this route have undergone the action of
fire.*

*Kirk-Andrews, Longtown, Westlington,
Carlisle.*—Sands and quartzose sandstones of
an ochreous red colour, limestones;—fine
cultivation in the neighbourhood of Carlisle
—large and excellent ploughs—a number of
kilns for making lime, which is used as
manure, not only for the meadows, but also
for the corn-land; small heaps of lime being
left to slack in the air, and afterwards spread
over the land which is immediately ploughed
over.

From a post-house before reaching Carlisle, a view is obtained of the Solway Firth,
which forms the western separation between
Scotland and England.

*Hairibee, Carleton, Low Hesket, High
Hesket.*—The same materials as above, that

---

* [In addition to the remark in the note on p. 261,
it may be stated that all the "lavas" mentioned in this
paragraph are Silurian sedimentary strata which do not show
any trace of "the action of fire." But on the hill-
tops above Mosspaul and Langholm true lavas and volcanic
vents of Lower Carboniferous age do occur. These,
however, are too distant to be detected from the road
down Ewesdale.]

is to say, sands, red quartzose sandstones, and limestones.

Penrith.—At a mile from this town, in descending the hill, we come upon large rounded blocks of basaltic lava, among blocks of granite, which are equally rounded.

Eamont Bridge, Cliston, Thrimby. — Blocks of reddish granite of a considerable size, with some rounded boulders of basalt, both lying upon beds of limestone.

Shap, Hausefoot. — Here the hills of tabular and fissile lavas re-appear.* Some of the rocks resemble those of Mont-Mezen, in Velay, which I have described in my *Minéralogie des Volcans.*

Kendal, Sizergh. — The same volcanic appearances.

Heversham, Milthorpe, Holme, Burton, Dure-Bridge, Carnforth, Bolton, Slyne, Lancaster.—This road runs for the most part through a calcareous country; rounded blocks of basalt are, however, sometimes observable scattered over the fields. The country is in general rich in pasture. The

* [The travellers were now entering upon the Silurian series of grits, greywackes and shales of the Lake district, similar in general characters to those of the south of Scotland which had also been mistaken for lavas.]

meadows are manured with a mixture of lime, stable-dung, and common earth, which forms an excellent compost.

From Lancaster we proceeded to Manchester.

CHAPTER XVI

*Manchester.—Doctor Henry; his Cabinet.
—Cotton Manufactures.—Messrs Thomas
and Benjamin Potter.—Charles Taylor*

IT was late when we arrived at Manchester. As I had letters of introduction to
Dr White and Dr Percival, I wrote to
them next morning, requesting to know
at what hour it would be convenient for
them to see me; but they were both going
to visit patients at the distance of some
miles from the town. As the matter was
pressing, Dr Percival sent a young German
of his acquaintance to express his regret
at not being able to see us; and Dr White
engaged his friend, Dr Henry, to wait
upon us in his stead. The latter devotes
himself successfully to the study of chemistry,
and has translated the works of Lavoisier
into English. These two gentlemen had
the complaisance to offer every service in
their power, and to shew us whatever was

remarkable in the town. They loaded us with every sort of politeness and attention, and hardly ever left us during the stay we made in Manchester.

Manchester is a large town; it contains between thirty-six and forty thousand inhabitants; but if we include all the people connected with a crowd of manufactures spread over the neighbourhood to a distance of three or four miles, it might be ranked among the chief towns of the second order.

The old cathedral is large and well built. We saw also some other monuments of this kind which were not uninteresting: but the cotton-mills which have created the wealth of this town, were an object that could not but excite our curiosity.

Nevertheless, despite the desire of our kind conductors to oblige us, we found it impossible to see any thing of the kind. Every attempt would have been vain; for the vigilance of the manufacturers had been redoubled, since they were persuaded that a French colonel, who had come to the town some time before, had wanted to procure plans of these machines in order to have them constructed in France.

Since then no stranger, and not even a citizen of the town who might be intelligent and observant, could have access to this kind of manufactures.*

The largest of these cotton-mills are moved by water: they spin the cotton to so much perfection, and with so much economy, that those who first erected them have made great fortunes. Arkwright, who invented them, was merely a barber in the town of Manchester, and all the more credit is no doubt due to him. He had the good sense to turn his discovery to profit, by entering into partnership with manufacturers whom he enriched, while at the same time he made a great fortune himself.

If I had not an opportunity of seeing the cotton-mills, they at least showed me with much politeness the largest warehouses

---

* At this time the machine for carding cotton had already been carried to France and was used there. Not long after, the mills were introduced by an intelligent Englishman, who disputed the merit of the invention with Arkwright. These ingenious machines are now erected in several departments where they are actively at work, until caprice and fashion return to the use of silk, that beautiful and sumptuous production of France, which employed such a number of hands and yielded revenues so immense.

of manufactured goods in velvet, in quilt-
ing, in dimity, etc., they unfolded the
finest pieces before me, and they even
invited me to take a look at the patterns.
We talked with some highly intelligent
men about the chemical processes connected
with the colours, and we received every
attention from Messrs Thomas and Benjamin Potter, as well as from Mr Taylor,
and I am desirous of expressing here my
gratitude.

I was doubtless indebted to Dr Henry
for all the politeness shewn to me in Manchester.   I wish that one day the opportunity may come to me of receiving him in like
manner in France.   This learned translator
of the works of Lavoisier studies chemistry
more than natural history.   At his house,
however, I saw some stones and minerals;
but what gave me most pleasure in his
collection, which is not large, was a fine
fossil femur of the unknown animal from
the Ohio, which is perhaps only a peculiar
and extinct species of elephant.   This femur,
which is in the most perfect preservation,
weighs forty pounds.

CHAPTER XVII

Departure from Manchester.—Buxton; its Mineral Waters; fine Baths, constructed on a Plan of Carr, at the Expense of the Duke of Devonshire, the Proprietor of the Waters.—Dr Pearson.—Manufacture of Vases and other Articles in Fluor Spar of different Colours.—Cave of Poole's Hole.—Toadstone composed of a basis of Trap, interspersed with grains of Calcareous Spar, presenting prismatic contraction as in basalt, though not the work of fire as the latter is.

WE were received in a very polite manner, as I have already observed, by Dr Henry, and by those to whom he was good enough to introduce us. But we were less kindly treated by the landlord of the Bull's Head inn, where we had put up. For two sorry dinners he charged us no less than seventeen shillings a-piece, with three shillings additional for his servants; and

exclusive of the bill for our own domestics.
Poor foreigners have nothing better to do in
such a case than to pay. Travellers are as
liable to this sort of exaction in Italy, in
Germany, and in France, as in England;
but this is not general and is absolutely alien
to the national character. It is the doing of
a few individuals only, neither very delicate
nor very just, who calculate badly what is
for their own interest; seeing that they soon
bring discredit on themselves and on the
houses they keep. It is a very difficult
matter to devise good regulations of the
police upon this subject. One thing well
known to those who are in the habit of
travelling, is that the inns where one is
worst lodged are always those where one has
most to pay. Until some remedy has been
found for this abuse, travellers should keep a
separate purse for these inn-keeping thieves,
just as is done in England for the *gentlemen*
who rob on the highways, to whom those
who venture to travel at late hours give,
without fear or danger, the remuneration
which is meant for them. I must say for my
part, however, that I passed through England
and Scotland twice, and by different routes,

without meeting any of these *gentlemen*; and that I experienced no extortion but at two places, *Dun's Hotel* in Edinburgh, and the *Bull's Head* in Manchester.

From Manchester to Buxton is twenty-four miles; but the road through Derbyshire is neither agreeable nor commodious. Sometimes it crosses over stony hills, sometimes narrow, wet and muddy valleys, and though the turnpikes are numerous and dear, the roads are in a pretty bad condition. They are in general, however, well supplied with post-horses. We left Manchester at seven in the morning, and did not reach Buxton till two in the afternoon.

Buxton is notable for its mineral waters, which attract a considerable number of visitors in the fine season. Buxton, however, is situated in the midst of the most dismal and cheerless country that I know. Its waters may be excellent, but most certainly the air that one breathes is impregnated with sorrow and melancholy. The houses, almost all uniform, but solidly built, look like hospitals or rather monkish buildings. An imposing fine erection in a grand and beautiful style of architecture, which is seen at the bottom of the place, and is appropriated to the baths, might be taken for the palace of the abbot.

We had letters of introduction for Dr Pearson, a London physician on the staff of the baths, who comes to spend generally six months of the year at Buxton. Fortunately he was still there, though it was now pretty late in the season. As he was well acquainted with the country, and had published an analysis of the waters, in which he mentions the stones and earths that form the soil of Buxton, he kindly and readily offered to conduct us to the most remarkable mines and caverns, the famous beds of limestone, traversed, according to him, by several currents of lava.

Dr Pearson is very intimate with Whitehurst, and has adopted his opinion respecting the beds of toadstone, which he regards as the product of the subterranean fires.* We fixed a day for going with him to see these supposed remains of volcanos in a country, where, on the contrary, everything indicates the agency of water. He had the goodness,

---

* *Observations and experiments on Buxton Waters, &c. By Doctor Pearson. London.*

in the mean time, to accompany us to the
shops of several workers in stone, who cut,
turn, and polish the fine Derbyshire fluor
or phosphoric spars * of different colours,
as well as gypseous alabasters, and some
marbles.

Workers in Fluor or Phosphoric Spar

Several workers in this line have settled at
Buxton on account of the numerous and, in
general, rich visitors who come for the waters,
and whose fancy or taste inclines them to
purchase their productions.   The fluor spars
are turned into small hollow or solid vases,
columns, eggs, pears, and watches ; they are
cut into pyramids, pedestals, &c.   As the
colours are beautiful and varied, and the
stone takes a brilliant polish, this material
was in great demand before abundant veins
of it were found.   Since then, the increase in
the number of workers and the consequent
competition among them, have greatly
lowered the price of these articles of
ornament.   It is rare to find among the
stone-turners of Buxton men who shew a
taste for beautiful forms.   These merchants

* The *Fluas calcareus* of the new chemical nomenclature.

have signs above their shop-doors with their names and the designation of " Petrifaction-Works."

One Noel seemed to me to be the most intelligent of them. He is quite comfortable, having succeeded well in this business. He is bringing up to the same trade a daughter and a son, who are already as well skilled in it as himself, though the boy is only eight, and the girl nine years of age. It was at his shop that I saw the best turned vases.

Samuel Cooper has the best-stocked shop; but he is dearer than the others. John Evans and Mottershed, are two other workers who have pretty good assortments.

It is necessary to be closely on the outlook for many little trickeries which they make use of to repair the accidents that happen to their stones, and consequently to deceive the purchasers.

They introduce, for instance, into the accidental cavities, or fractures, which they are dextrous enough at mending, lead as found in nature, that is, as it comes from the mine (galena). They then polish it; and do not fail to assure the purchaser that this lead is natural to the spar, and makes it more interesting.

I remarked also, that to give greater brilliance to their productions, they have always water at hand to plunge them into, on the pretence that it was only to wash off the dust. The water has a singular effect in enlivening the colours, the polish, and the semi-transparency of the stone.

The fluor spar, which is worked at Buxton, comes from the lead mines of Castleton, about ten miles distant. The only stones of value found in the hills around Buxton, are a very fine gypseous, white, semi-transparent alabaster, which is made into vases and pedestals, a black marble emitting a somewhat bituminous smell on being rubbed, and a yellowish calcareous spar, both of which are applied to the same purpose.

The Bath-house

This superb edifice better deserves the name of a palace [Palais des Thermes] than that of a place for bathing. It is a vast fabric, in the shape of a rotunda, ornamented all round on the outside with large pilasters which support a rich cornice crowned with a balustrade in stone.

This building, in addition to the space occupied by the baths, is intended to accommodate two hundred guests, exclusively of rooms for their domestics, for the persons employed about the baths and wells, and for the different innkeepers and caterers who have to provide for so numerous an assemblage, and who are the principal tenants of the whole structure. There are also coffee-rooms, gaming-rooms, and ball-rooms.

The whole of this fabric, erected at the expense of the Duke of Devonshire, after the plans and under the superintendence of the architect Carr, is in a good style of architecture, uniting to an air of grandeur a feeling of taste, which does honour to the talents of that able artist, whom I had the pleasure of seeing, and who was kind enough to conduct me through every part of it.

The baths are disposed with the same judgment as the rest of the establishment. There are common and private ones for the women; those for the men are in a separate quarter and possess the same conveniences.

Baths have also been constructed for the poor.

The mineral waters for drinking run into a large cistern of white marble, placed under a pretty little temple finely executed in antique style.

The waters of Buxton are rather tepid than warm; as they do not raise the mercury in Fahrenheit's thermometer above eighty-two degrees. Dr Pearson, who has analyzed them, says, that the air which is disengaged from them in great abundance, does not contain any fixed air, but atmospheric mephitis, or the *azotic gas* of the new nomenclature.* This is a very remarkable fact.

The village of Buxton is not very considerable, and the greater part of the houses belong to the Duke of Devonshire. They are let to inn-keepers, and yield him a great revenue; a single one of these taverns, that

* The term *azote*, derived from the Greek, signifies *privation of life*; but, as other gases produce the same effect, the word is improper. I do not attack the principles of pneumatic chemistry, but its language. If this malady, as hideous for the language of Racine, Fenelon, Buffon, Voltaire, &c. as leprosy is for the body of a beautiful woman, gains ground in the other sciences, we shall soon realize the symbol of the tower of Babel, and the confusion of language will bring with it the confusion of ideas: and thus we shall reach barbarism.

nearest to the baths, is rented at twelve hundred pounds sterling; and, I was assured, that the baths alone bring him at least thirty-six thousand French livres.

To draw a greater number to the houses belonging to himself, the Duke of Devonshire adopted a plan which has succeeded. Those who are lodged in any of them are charged only sixpence a-day, for the waters, whilst such as have apartments elsewhere pay a shilling a-day.

Lithology of the neighbourhood of Buxton

Buxton is surrounded with many small hills, placed close to each other, the highest of them not exceeding some six hundred and fifty feet in height. A little stream, the Wye, takes its rise at a short distance from the fine edifice of the new baths; and soon after plunges into a pretty deep ravine, and runs between two low hills closely adjoining.

The hills on the road from Buxton to Manchester are composed of beds of sandstone, in some parts hard and in others soft and friable, and sometimes laminated. The hard sandstone is here called *greet*, *grit*, or freestone, and sometimes *millstone-grit*; the

laminated kind is termed *slate*. These
sandstones are disposed sometimes in large
horizontal banks, and at other times in
laminated strata ; they are usually white or
reddish.

Near these, and almost opposite to them,
are hills entirely formed of limestone in
horizontal beds in some places, and in a
continuous uninterrupted mass in others.
These masses are sometimes divided by perpendicular or diagonal fissures.

This limestone is hard, almost sparry,
and yields when burnt a lime of the very
best quality. It is used also as a wrought
stone for making door-frames, chimney-
pieces, pedestals, and other works. Its
colour is, in some quarries, whitish or greyish,
and in others black. The latter is employed
for the same purposes as marble ; and I
have almost never found any marine bodies
in it ; it gives off a very disagreeable smell
on being rubbed ; it may be regarded as a
kind of swinestone (pierreporc).

The greystone, on the contrary, contains
madrepores, entrochi, and other marine
petrified-bodies ; likewise siliceous nodules
which are full of entrochites. It does not

possess the same hardness throughout; its soft part emits a disagreeable smell on being rubbed, not so strong, indeed, as that of the black kind, but nevertheless very appreciable. The harder pieces of the greystone are used for several domestic purposes. I shall return to these limestones, which are remarkable for the circumstance of being interrupted with alternate beds of toadstone.

In that part of Derbyshire called the *Peak*, in which are situated Buxton, Matlock, Wirksworth, Middleton, Masson, Cromford, Winster, Castleton, Eyam, &c. there are found black argillaceous shale, more or less hard, and sometimes resembling slate, black and grey clay-ironstone (argile martiale), more or less hard, which is not put to any use, and red or grey marl, a brownish kind of marl of a very fine grain, charged with calcareous particles, which is used like tripoli for polishing tin, copper, crystal, &c. and is called rottenstone. Further, in this tract there are mines of coal, in full working, usually not very deep, having for roof a black argillaceous shale, full of prints of ferns, the greater part of

which seem to be foreign. Here are likewise found pyrites, black marble, grey
marble, soft limestone, sandstone, compact gypsum, striated gypsum, fluor-spar
amethyst-coloured, yellow, red, grey, or
white; and a heavy very fine white earth,
as easily cut as chalk, said to abound in
gypseous earth, and indifferently known
here as *caulk*, *calk*, *cawk*, *kewel*, and *keble*
according to the places in which it occurs.
This substance is one of the most common
gangues in the mines of Derbyshire, and is
frequently seen adhering to fluor spar.*
Moreover the district furnishes doubly-
terminated rock crystal, heavy spar, and
also opaque fluor-spar, in detached cubical
crystals, (these three substances being found
on Diamond Hill, about three miles from
Buxton); calcareous tufa, with plants incrusted in it, manganese in kidney-shaped
lumps, lead, copper, calamine, and blende;
the thermal waters of Matlock Old Bath,

* It is thought that the *caulk* makes the regulus of antimony
more ductile, and of a closer grain. It is used at Birmingham
in the manufacture of brass. Some persons imagine that it is
for the purpose of making moulds, but of this there is no
certainty, the greater part of the processes being concealed in
the English manufactories.

Matlock New Bath, and Buxton; the acidulated mineral waters of Quarn, or Quarnden, an intermitting spring at Tideswell, and a number of natural grottos or caverns, several of which are very deep, such as those of Castleton, Poole's-hole, Elden-hole, Hosen's-hole, Burmforth-hole, and Lathkill. It is very remarkable that pretty copious streams of water flow through almost all of these caverns.

Such is a brief sketch of the principal objects observable in the Peak of Derbyshire; their astonishing variety within so small a compass, cannot but amaze the naturalists most versed in geology. Accordingly Mr Ferber, so well known for his "*Letters on the Mineralogy of Italy*," in the preface to a pamphlet which he has published on Derbyshire, acknowledges that, without the aid of Mr Whitehurst, to whom the ground was familiar, and Mr Burdett, who has published a fine map of Derbyshire, he would have been completely bewildered in so extraordinary a country.

"I sincerely acknowledge," says he, "that without the assistance of these two persons, I should have often found it diffi-

cult to explain a great number of phenomena which were new to me. I was till
then acquainted only with homogeneous
mountains; and among all the stratified
mountains which I had examined, and of
which the internal structure was perfectly
known to me from visiting the mines, not
one furnished any instance to be compared
with what I saw for the first time in
Derbyshire. The great diversity of the
strata and their frequently fantastic disposition, which I had never seen any where
else, very often embarrassed me; and I am
persuaded that the ablest mineralogists will
experience the same difficulty."

The Cave of Poole's-Hole

Mr Ferber, in his " *Essay on the Oryctography of Derbyshire*," does no more than
slightly mention the cave of Poole's-hole.
" This cave," says he, " which is at a little
distance from Buxton, abounds in stalactites;
it is said to be half an English league in
length, and a very noisy stream runs
through it."

The following is a more minute description :—Poole's-Hole is about a mile from

Buxton; and its entrance is at the foot of a large limestone hill, where the rock, everywhere bare, presents a number of open quarries and many kilns, where the stone, which is of excellent quality, is burnt into lime. This limestone contains a number of entrochi and other marine bodies, converted for the most part into calcareous spar.

Dr Pearson was so obliging as to accompany us to the cave. Scarcely had we arrived at the mouth of it, which is narrow and oblong, when several women hastened to join us; some to sell us bad stalactites and bits of shining calcareous spar; others to bring lights, and to offer their services in the capacity of guides. We accepted their offers, and entered the cave. Here, as in almost all natural caverns, there are narrow galleries, and others much broader, here tortuous, there in straight lines, while at intervals they expand into spacious and lofty chambers. We were made to stop in front of a very large stalagmite, which is only an irregular and confused accumulation of calcareous spar deposited by the drippings. The natives call it *Poole's Chair*. They have heard it so named by their fathers, those who come after them will give it the same name; and though this sparry mass has no more resemblance to a chair than it has to a horse, the power of imagination and habit will maintain this absurdity, and those good folks will always like to think that they see what they do not see. Alas! it is the same with many other things in this world.

On penetrating into a deeper recess, our conductors did not fail to tell us that this was *Poole's Chamber*; and a little farther they pointed out to us *Poole's Table.**

The cave is about two thousand and eighty-five English feet in length at the most, even including certain inconvenient passages. A thread of water, which perhaps becomes a more considerable stream in rainy seasons, runs through the whole length of the galleries, and makes the walk rather uncomfortable. On the whole, this cave offers nothing of interest; one sees only a few bad and mutilated stalactites, and these not in any great quantity. The roof of one of the galleries has fallen in, and whether from the

* Credulity has always been pleased to see something marvellous in these subterranean caverns far from the light of day. There is still shewn in the cave of Sassenage, near Grenoble, the famous table of the fairy Mellusina.

excessive weight of the super-incumbent mass, or from the shocks of earthquakes, huge blocks of rock have been detached, which by their number and size block up certain passages.

On coming out of the cave, we were anew besieged by women, who offered us for sale poor fragments of crystallized calcareous spar, which they seemed to think of much value.

We then visited the numerous quarries of limestone opened on all parts of the hill, above the grotto of Poole's-Hole. More than a hundred families are engaged, from father to son, in working the stone and making lime. The consumption of this article must here be enormous, and the markets for it considerable, for lime-kilns are seen smoking on every side.

I looked in vain for the habitations and houses of so many workmen and their large families, without being able to detect the smallest cot, when I all at once discovered that the whole tribe, like moles, lived under ground. This comparison is just; for not one of them was lodged in a house, or even in an opening of the rock. They had

preferred to scoop out their dwellings among
heaps of cinders and lime-refuse, which
formed so many little monticules, or, if one
may so say, mole-hills.

The workmen have hollowed out for
themselves subterranean habitations in these
old piles of rubbish, which has been consolidated by rain into a kind of compact
cement, which is now impenetrable to the
water.   As the excavation of this material is
easy, these families have known how to
obtain protection from cold and wet, by
burying themselves, and even backing up
their dwellings against the lime-kilns which
thus warm them.

Most of these habitations have three or
four rooms, almost all of a round shape, for
the purpose of greater solidity.   They are
lighted from the side, when the position
allows it, or merely from the chimney, which
is a round hole in the roof to let the smoke
out.   Openings like dormer-windows are
also made by the door of the place to admit a
little light.   Thus, at the dinner hour, when
the workmen have gone into their holes, and
a multitude of little columns of smoke are
seen coming out of the earth, one might

suppose it to be a large village in Lapland. I had much pleasure in paying a visit to this kind of troglodytes. They cannot get rich at this trade, since they cannot afford themselves the smallest comfort, and though in the midst of stones and lime, they have not yet been able to build a wall.

Of the stones known in Derbyshire by the name of Toadstone

As I had here reached the source of those stones, famous among English naturalists, and the same which served as the foundation of Mr Whitehurst's theoretical system, it was my wish to examine them with the most scrupulous attention. I had the advantage of being accompanied by a mineralogist of Mr Whitehurst's school who was well acquainted with the localities, and who, like him, appeared to be convinced that the toadstones were real lavas. I was under promise, besides, to give an account of my researches to Mr Whitehurst himself. Being provided with his book, and Mr Ferber's essay on the Oryctography of Derbyshire, it only remained for me, as far as the toadstone was concerned, to make myself acquainted with the vocabulary of the miners. The names which they have given to this stone vary according to its colour, its hardness, the bodies found in it, and the disposition of its beds. These names are not scientific, it is true; but the miners are accustomed to them, and perfectly understand them, being derived from their native language; and most assuredly they will never think of changing them.

Toadstone has been so called because its ground colour, which is very dark brown, sometimes black, is dotted all over with globules of white calcareous spar [giving a fancied resemblance to the back of a toad]. These white specks are, in general, pretty uniform, and sometimes protuberant. A stone in every respect similar is found in the bed of the torrent of the Drac, near Grenoble, where it has been carried down from the High Alps of Champsaur. Its spotted appearance has led to its being called Variolite of the Drac. But this term already affixed to a stone of a different sort, ought to be rejected, in order to prevent mistakes.

The word toadstone is much in use at Buxton, Matlock and Winster, because stone

of this character is more common there than in
the other lead or copper mines of Derbyshire.

At Tideswell, the same stone, but with
few or no globules of calcareous spar, and
disposed in thick beds, which alternate with
layers of limestone, bears the name of
channel.

At Ashover, in the Gregory mine, being
of a blacker kind and not so hard, it is called
black clay.

At Castleton, the miners have given to a
greenish variety of this stone, which falls
into earth on exposure to the air, the singular
name of *cat-dirt*.

Mr Ferber, in his essay on the Oryctography of Derbyshire, pages 163 and 168,
mentions only the two designations of *toadstone* and *channel*; but he adds, that the
same stone bears in England the names of
dunstone and *black-stone*, and in Scotland,
that of *whinstone*. It is certain that the
Derbyshire miners know very well how to
apply the four different names which they
give it, according to its different modifications. This is not the case with respect to
the English term *black-stone*, which is
applicable to other stones of the same

colour, but of a very different nature, such as volcanic basalt, touchstone, black schorl, hornblende, certain fine grained black granites, &c. This denomination is, therefore, too general and vague.

That of *whinstone*, which the Scots employ, is no better. They give this name to every black stone, which is hard and rough to the touch. At Edinburgh, when I asked the workmen and naturalists of the country to shew me whinstone, some of them brought me a black hard stone of the same nature with what the Swedes call trap, which is not in the least volcanic; others presented me with a compact lava of the nature of basalt, and others a variety of black granite, which has undergone the action of subterranean fire without any alteration in its hardness, and which makes most excellent pavements. The case was nearly the same at Glasgow. This term, then, having, like the preceding, several acceptations, which tend to spread confusion of ideas, ought to be rejected. A fact worthy of remark, and which proves that the character of colour has been more attended to than any other, is, that in

Germany, where the science of mineralogy has long flourished, several hard black stones, of very different natures, are likewise denominated *black-stone* (schwartzstein).*

Mr Ferber, though profoundly versed in mineralogy, does not compare the toadstone of Derbyshire with any other stone, gives it no distinctive name, and says not a word about the opinion of Mr Whitehurst, who regards it as a volcanic product. He contents himself with stating, that "this stone has an argillaceous base more or less hard; some parts appearing to be merely an indurated clay, whilst others approach jasper in hardness; that it is interspersed with small grains or globules of calcareous spar, varying in form and size, but sometimes so minute that to the eye they seem to be lost in the black substance of the stone itself, whilst others are as large as a

* In France, as in other places, names have sometimes been given to mountains from the colour of their rocks; such as roche *maure* for roche noire (black rock), roque *brune* for roche brune (brown rock), *peire neir*, pierre noire (black stone), &c. But whenever I have shewn to intelligent peasants pieces of compact lava, or blocks or columns of basalt, and have asked them how they termed these stones, they all have said that they were *pierres mortes* [dead stones.]

pea, or even a bean." He adds that he had "tried the stone with acid, which dissolved with effervescence the portions of calcareous spar, without altering the substance of the stone itself; that, after the experiment, it was still hard enough to scratch glass, though it gave only a few slight sparks with the steel; and finally, that the substance of toadstone, when divested of its calcareous particles, seemed to him to be refractory before the blow-pipe, but that with the addition of salt of tartar he converted it into a black scoria, which would seem to indicate a siliceous principle, though it does not possess the hardness of siliceous stones."

To enable those naturalists, who have not been on the ground, to form a correct notion of this stone, and to follow me in the examination which I shall now present, it only remains that I should give them a preliminary description of the singular position which it occupies in the mountains of the Peak of Derbyshire. A part of the details I shall borrow from Mr Whitehurst and Mr Ferber, and add what I have seen myself.

*Of the different substances which precede
and accompany the toadstone of the Peak
of Derbyshire*

1. Quartzose-sandstone (greet, grit, freestone, millstone of Mr Whitehurst, from the
use to which it is applied in some places).
It varies in colour, grain, hardness, and the
thickness of its strata. Mr Whitehurst says,
that the bank is 120 English feet in thickness, that this sandstone is made up of
rounded grains of quartz and little fragments
of the same substance, wherein the rough-
nesses of the fractured surfaces are still quite
visible.*

2. Black argillaceous shale of the nature
of slate or shiver. Its thickness, according
to Mr Ferber, who measured it in the mine
of Yatestoop near Winster, is from a hundred
and forty to a hundred and fifty feet;†
and according to Mr Whitehurst who probably measured it in another place, about
a hundred and twenty yards.‡ As it is
well to be familiar with the nomenclature of

* "Inquiry into the original State and Formation of the
Earth," p. 147.
      † Page 160.   ‡ Page 148.

the miners, when one is on the ground, it may be stated here, that they have given several names to the toadstone according to its greater or less hardness, or as it is more or less penetrable. Hence they call it *shale*, *hard-beds*, *penny-shale*, and *black-beds*.

In some places, among others near Winster, where the high-way is opened through beds of this shale, which are completely visible, there are found below it large pieces of black lime-stone, which when rubbed with iron, give out a strong smell of burnt horn.

3. First Limestone is black, very hard and used as marble near Dashford; giving an offensive smell when rubbed; in some places without marine bodies; in others, abounding in anomias and various bivalves; sometimes contains lumps of silex [chert], and sub-divided at intervals with thin seams of a kind of slate. The thickness of this bank varies from thirty-five to fifty feet.

4. First bed of Toadstone. The thickness of this bed varies greatly: in some places it is fourteen, and in others sixteen feet, and at Tideswell it has been penetrated to the depth of a hundred and sixty feet without

reaching its bottom; though, in the same mine, at the distance of a hundred toises only from this spot, it is no more than forty feet; and three hundred toises farther, only three feet. This would seem to shew that the material which has formed this stone has accumulated here rather in one large deposit than in regular beds.

But let us hear Mr Whitehurst himself, in order to learn his very explicit opinion upon this stone: "It is a black substance, very hard; contains bladder-holes like the *scoria* of metals, or Iceland lava, and has the same chemical property of resisting acids. Some of its bladder-holes are filled with [calcareous] spar, others only in part, and others again are quite empty. This *stratum* is not laminated, but consists of one entire solid mass, and breaks alike in all directions. It does not produce any minerals or figured stones representing any part of the animal or vegetable creation.—Neither does it universally prevail, as the limestone strata do, nor is it like them equally thick; but in some instances varies in thickness from six feet to six hundred. It is likewise attended with other circumstances, which leave no room to doubt of its being as much a lava, as that which flows from Hecla, Vesuvius or Etna."

A second reason, which has induced the English naturalist to regard the toadstone as a true lava, posterior to the formation of the calcareous beds, is, that the vertical fissures sometimes observable in these beds, are filled with toadstone; which necessarily implies the pre-existence of these fissures, and consequently of the calcareous strata.

5. Second or Grey Limestone: about thirty-three toises thick, grey and enclosing many petrified marine bodies, including species of *Chama* of a very large size, madrepores, &c. The stone is not equally hard throughout; the softer part which serves for making lime, emits a disagreeable smell on being rubbed with iron; the hard part can be hewn, polished and put to the same purposes as marble. Some parts of it are penetrated with siliceous matter enclosing entrochi.

6. Second Bed of Toadstone, forty-six feet thick. Mr Ferber says, " that in the mine of Hubberdale, this stone has lost its usual hardness to such a degree that it perfectly resembles a soft clay." But this

alteration, which occurs in some parts of the mines, is merely local; and the toadstone of this second bed, as Mr Whitehurst has well observed, is more solid than that of the first. A circumstance worthy of attention is, that it contains not a cavity, and consequently not a globule of calcareous spar.

7. Third Limestone; grey, like the second, and, according to Mr Ferber, seventy yards thick. Mr Whitehurst says, that it contains fewer petrifactions than the preceding ones, and that it is thirty fathoms thick.

8. Third bed of Toadstone, resembling the second; twenty-two feet thick. " In the mine of Hubberdale," says Mr Ferber, " this stone has the consistence of a tender and soft greenish clay filled with small nodules of black clay, and white calcareous spar in veins: it is called *channel*."

9. Fourth Limestone; grey, like the preceding, but a little whiter; thickness unknown; it has been pierced in some mines to a depth of forty fathoms, without reaching its bottom. It is not therefore known whether this fourth calcareous bed, which is so thick, is again succeeded by toadstone.

The direction of the metalliferous veins is in general very regular; the salband* is distinct and well-marked ; its breadth varies, being sometimes several feet, and often several toises.

But what is truly an extraordinary phenomenon in mineralogy, is, that the veins, which are very rich in the four limestones, disappear, as they approach the beds of toadstone, which alternate with the limestone, so that no matter how thick the toadstone may be, its whole mass must be pierced without any indication or the slightest trace of ore, until the limestone is reached, where the ore never fails to reappear. Thus, for example, when a vein has been worked out in the first bed, that is, in the black limestone, the ore is at once lost on reaching the toadstone, and does not again make its appearance till the bed of toadstone has been entirely passed through. "This phenomenon," Mr Ferber justly says, "is unquestionably one of the rarest and most singular of its kind; and to account for it is no less difficult. Another

[The name given to the soft earthy material frequently found along one or both of the walls of a metalliferous vein ; *flucan* of Cornish miners.]

peculiarity in the beds of toadstone is, that
this curious substance so completely separates
the different strata of limestone, that the
flooding of a gallery in the first bed nowise
disturbs the work in the second, and the
miners may be dry in a lower gallery, whilst
all the galleries above it may be drowned
out."

Great subsidences and violent disruptions,
at periods lost in the night of time, have
depressed the banks of stone which have
sunk to a great depth in some places, while
from the same cause they have been upraised
in others. New adventitious earths and
extraneous materials, transported by other
revolutions, have filled up many of these
depressions, and have thus partly concealed
the first framework of this astonishing
country; but the deep excavations which
the miners have made through a great extent
of the high plateau of the Peak, have
furnished the means of obtaining an exact
knowledge of the topography of Derbyshire,
of which I have here given a slight sketch,
and of which much more detailed explana-
tions will be found in the work of Mr
Whitehurst, who has had engraved, from

very accurate drawings, plans and profiles of the most interesting sections in these mountains.

Much impressed with the subject, and having several times before reaching Buxton stopped at the foot of some crags where the toadstone was exposed, I asked Dr Pearson, whether, in his work on the mineral waters of Buxton, he had particularized any bed of toadstone which we might visit together, assuring him, that I would be all the more pleased to gain my information with him, as nothing that I had yet seen in that part of Derbyshire bore to my mind the least trace of a volcano, and this same toadstone which, with Dr Whitehurst, he looked upon as a true lava, was found on the High Alps of Champsaur in Dauphiny; that precisely the same varieties of toadstone were to be seen there, with globules of calcareous spar, with empty cells, devoid of spar, often without pores, sometimes hard and compact, at other times soft, decomposing and changing colour, and further that it had been sometimes found divided into small prisms, which led me to think that we should probably meet with it here in the same condition.

I mentioned also, that M. de Lamanon, a
very estimable naturalist, who fell a victim
to his ardour for natural history, in the
voyage round the world with La Peyrouse,
had written a pamphlet to demonstrate that
the stone of Champsaur was a product of
the subterranean fires, and that he considered
himself the discoverer of an extinct volcano
in the Alps; though very able naturalists
had affirmed there exists not a trace of a
volcano among these high mountains. But
I told him at the same time, that M. de
Lamanon, whose opinion I combated, relin-
quished his error, and suppressed the whole
impression of his work, only reserving
twelve copies, at the end of each of which
he had the honourable frankness to print his
recantation; that these few were sent to
those who had first opposed his views, as an
acknowledgment that the stones which he
had taken for lavas were merely traps, and
that he had the goodness to send me one
among the rest.*

* See my essay on trap-rocks (1788, in 12mo. p. 31),
in which I have traced the itinerary from Grenoble to the
mountain of Chaillot-le-Vieil, in the Alps of Champsaur, with a
description of all the varieties of trap, in every respect like those
of Derbyshire, in a much higher country; for the mountain

Excursion in the vicinity of Buxton with Dr Pearson.—Bed of Toadstone of which he has spoken in his Book on the Mineral Waters.—Small isle in the River Wye, formed entirely of Toadstone divided into Prisms

" Let us set off then," said Dr Pearson, " I shall be very happy to shew you the bed of toadstone that I have mentioned, and you will tell me what you think of it." We descended into the ravine which serves as the bed of the little river Wye ; this stream, if we may judge by the channel which it has excavated, must become a torrent in rainy weather. We kept up its course for about a mile as far as a corn-mill.

Before getting there, particularly on the right bank, and immediately under the vegetable soil, we noticed some beds of black shale of variable thickness and splitting up under the action of the air. This shale is sometimes covered with a slight efflorescence of iron-sulphate. It is two or three feet

Peyre-Niere (Pierre noire) or the Haut-Puy, one thousand four hundred toises above the level of the sea, is crowned with real toadstones, that is to say, with traps.

thick, disappears at intervals under the vegetable soil, and is lost to sight as it approaches the mill.

Here the nature of the ground suddenly changes; the valley diminishing into a narrow defile between two limestone hills, where the mill has been erected; the constriction of the channel having no doubt suggested this as a convenient situation.

The limestone is grey in colour, and its strata dip from both sides towards the bed of the stream; but so vigorous is vegetation in this moist region that, except in a few places, the rocks are completly covered with mosses, lichens and peaty earth, and obstructed with creeping plants.

The cart-road leading to the mill runs along a natural causeway formed by the rocks, which are entirely bare in this deep hollow. A little above this mill the road is crossed by a bed of toadstone several feet thick, the black colour of which forms a striking contrast with the grey tint of the limestone.

This is the bed of toadstone which Dr Pearson has drawn as alternating with limestone banks. But on examining it attentively,

I pointed out to this naturalist, how difficult it was to determine whether we had here a real bed rather than a kind of vein; for the great subsidences which the two limestone hills have undergone and the vegetable soil which for the most part covers them, scarcely permit one to pronounce with certainty as to the exact and primitive disposition of the limestone banks.

When the toadstone is examined in those places where it is exposed, it appears rather to intersect transversely the limestone beds than to follow them, a fact which, if sufficiently ascertained, would destroy the Doctor's hypothesis founded on the stratification of the calcareous beds alternating with the toadstone.

When we carefully examine this little valley or rather enclosure formed by the subsidence of the two limestone hills, and through which the Wye now flows, we are led to believe that a revolution, subsequent to that which gave rise to such great displacements, has filled up, with deposits and secondary alluvia, the disruptions, the cavities and the fissures that were produced by the concussion and the fall of these enormous masses.

I submitted these reflections to Dr Pearson, stating at the same time, that my conjectures would have still greater probability if we should find, at the bottom of the valley, the toadstone in mass above the limestone.

Whilst making these observations, I cast my eyes on a small oblong isle in the very centre of the place in question. " Let us see," said I, " whether that kind of natural barrier, which by its resistance, has made the river divide into two branches, consists of the same stone as that of the neighbouring hills."

Dr Pearson replied that he had not carried his enquiries so far as that. We then repaired to the small isle, which is hardly more than a hundred paces long, by ten to twelve broad, though it must have been much more considerable before the waters had worn it away. It rises only a few feet above the stream.

With equal pleasure and surprise we found that it was entirely composed of a blackish brown toadstone, full of globules of calcareous spar in some parts, while in others these inclusions were fewer, or were entirely absent. But what is remarkable is, that the first or uppermost bed of toadstone, which is two feet and a half thick, presents places where the stone is disposed in prisms, giving a most exact representation of a small basaltic causeway. It is still further astonishing, that as basalts are sometimes found in spheroidal forms beside the prisms, and as the prisms themselves, occasionally losing their angles, give rise to ball-like lumps, so here, too, the toadstone assumes all these forms. The balls exfoliate in concentric coats, as do the spheroidal basalts.*

---

* "Trap," says the celebrated Bergmann, "is found, though rarely, in the form of triangular prisms. It sometimes has the appearance of immense columns, such, for instance, as the *Traclestemar* at the foot of the mountain of *Hunneberg*, opposite to *Bragnum*, which have detached themselves from the rest of the mass. The first time I saw them they formed an angle of eight degrees from the perpendicular. In almost all the mountains of Westrogothia that have a stratified form, the upper bed consists of trap. It is interesting to observe, that this bed always lies on a black aluminous slate. Is it possible then that this matter should have been in a state of fusion without the slightest diminution, the slate underneath losing, even at the point of contact, none of its blackness, which it certainly does when put into a common fire? We have a still finer trap, which generally runs in veins, and is frequently found in very ancient mountains, wherein not the faintest marks of subterranean fire can be discovered."

Lettres de Bergmann a Troil, page 448 (*Lettres sur l'Islande*).

These prisms and balls are in a state of incipient decomposition; the colour of the stone is brown, sometimes yellowish iron-grey; its base (pâte) is mixed in some parts with a multitude of globules of calcareous spar, the colour of which is frequently stained with the tints produced by the decomposition of the toadstone. This stone, thus disposed in prisms and balls, rests on a bed of friable and gravelly material, which is in reality nothing but toadstone reduced by decomposition to the form of a sandy earth.

It must be acknowledged that nothing can be more volcanic in appearance than this little isle of toadstone; for a vein of this material, which has some resemblance to a current of lava, can here be seen to cross the limestone which forms the bottom of the cart-road, and then to sink and lose itself in the Wye, so as to lead one to imagine that it has given rise to the isle, composed of a substance which not only has the colour and aspect of certain lavas that seem riddled with pores where the globules of calcareous spar have been destroyed, but is moreover dis-posed in prisms and balls. There is nothing, however, really volcanic, either here or in the neighbourhood.*

This serves to shew how useful in natural history are local descriptions made with exactitude, and how important it is, in certain circumstances, to have the opportunity of seeing objects in place.

The substance of toadstone is a compound of siliceous earth and argillaceous earth, with a small quantity of calcareous earth and of iron. The proportions of these materials

---

* [Notwithstanding the author's important contributions to the literature of volcanic rocks and his services in maintaining the volcanic nature of basalt, his reasoning in this passage is worthy of the most thoroughgoing Wernerian of his day. He recognised in the toadstone a number of features which closely resemble well-known structures of ordinary lavas, and which it might have been supposed that he would have been delighted to detect as evidence of former volcanic action. But for some reason which it is not easy to discover, he had made up his mind that toadstone must be a form of "trap," and like the most orthodox follower of Werner, he did not regard "trap" as an igneous, but as an aqueous rock. He gives no explanation why the apparently volcanic features of the rock should be set aside. After an expression of doubt as to the original relations of the limestone and toadstone, and a somewhat nebulous reference to the effects of revolutions and disruptions (which even if accurately stated would have no real bearing on the main question) he dismisses the discussion as to the history of toadstone with the oracular affirmation that "there is nothing really volcanic either here or in the neighbourhood."

In the first chapter of the first volume of this "Journey" he narrates his conversation with Whitehurst in London on the

differ according to the varieties of the toadstone. That of Derbyshire, which is the subject of our present enquiry, has been analysed by Dr Withering, who found that a hundred parts contain sixty-three parts of siliceous earth, fourteen of argillaceous earth, and seven of phlogisticated iron.

I myself, also, have analysed a piece of the same stone which was broken off from a part that had no calcareous particles ; the results which I obtained were a little different.

subject of toadstone, and mentions his promise that he would, on his return, pay a visit to Derbyshire, and report to this venerable cosmogonist what conclusion he had come to from seeing the rock on the ground. It is clear, however, that from the specimens which Whitehurst shewed him he expected to find that his friend had made a mistake. Whether he again saw the author of the " Inquiry into the original State and Formation of the Earth," or wrote to him from Buxton, as he said he would do, is not known. But if he communicated his views to Whitehurst, they made no change in the old man's published opinions, for two years later (1786) a second edition of the Enquiry was published in which the volcanic origin of the Derbyshire toadstone was asserted in the same emphatic words as in the first edition. The researches of recent years have completely vindicated Whitehurst's opinion. Not only do the toadstones include lavas, which were poured out over the floor of the Carboniferous Limestone sea, but also layers of volcanic ashes, and intrusive sills intercalated among the limestones, while even some of the volcanic vents have been recognised from which these various materials were ejected.]

From a hundred grains of it the produce was as follows:

Siliceous earth . .	54 grains.
Argillaceous earth . .	19
Aerated calcareous earth	8
Aerated magnesia . .	4
Iron	13
Lost	2
Total	100

In making other experiments on stones of the same kind, taken from different beds, I found the same constituent principles, but always with greater or less variation in the results; sometimes the quantity of iron, of calcareous earth or of argillaceous earth is increased or diminished. In a word, and to conclude this already too long and too wearisome discussion, the toadstone of Derbyshire has absolutely no connection with volcanoes, and is nothing else than the Trap of the Swedes.*

* If still further information should be desired respecting this stone, I may refer the reader to pages 7, 23, 31, 43, and 53, of the work which I have published on Trap-rocks; where he will find that the *argilla martialis indurata* of Cronstadt, the *schwartzstein* of the Germans, the *toadstone, channel, cat-dirt,* and *black clay*, of Derbyshire, the *whinstone* of the Scots,

Some may perhaps blame me for wishing to generalise too much the name given to a stone which is in the class of compounds, and which itself serves as the base of other stones. But I have never claimed that the name of trap should be given exclusively to rocks in the composition of which the material of trap abounds, in all cases where these rocks have some character that differentiates them. I have not, for example, ceased to use the terms porphyry, amygdaloid, variolite, etc., though trap be the base of all these stones.

I perfectly agree with my illustrious friend, M. de Saussure, who says that " when two fossils exhibit notable differences we are not to refrain from distinguishing them by different names, on the pretence that intermediate varieties are found which appear to unite them, by seeming to belong equally to the one and to the other." *

and the *variolite du Drac* of some Frenchmen, are nothing else than trap, more or less hard, and more or less altered. This stone also forms the base of the most of the porphyries, &c., &c. But I am of opinion that, while preserving the generic term Trap, we ought to add the different popular names which the miners or general custom have given to this stone on account of its various modifications.

* "Voyage Dans les Alpes," 4to., Tom. iv., p. 127.

It is in order not to depart from this principle, that, while always preserving the generic term *trap*, which must be respected, seeing that it has always been admitted by the mineralogists of the North, I call a stone trap when it has the hardness, the colour and the kind of homogeneity which is proper to itself, without the addition of any well defined foreign ingredients. But if it contains globules of calcareous spar, I would call it, with the miners of Derbyshire, *toadstone*. When the trap, however pure, has lost its hardness, and its original colour, particularly when the colour inclines towards green, and when at the same time I find this trap disposed in a kind of veins intersecting rocks of another nature, I would make no scruple to call it *cat-dirt*; and so on with the other known modifications as often as they are sufficiently distinct. This is the simplest way of coming to an agreement, and at the same time respecting the works and the memory of those savants who have cleared the paths of science before us and have been our first masters.

CHAPTER XVIII

*Castleton.—Description of a fine Cavern.—
Mines of Lead and Calamine, Veins of
Fluor Spar.—Lead found in Channel or
Cat-dirt*

WE rode from Buxton to Castleton, a
distance of twelve miles, on one of
the finest days of autumn, but at the same
time upon a road as detestable and fatiguing
as the worst roads in winter.

During our stay in this little place, which
is pleasantly situated among hills we saw
the different workers in fluor spar, and
visited the magnificent cavern called the
Devil's Bottom, near Castleton, as well as
several mines in its vicinity. The result of
my observations is as follows:

DESCRIPTION OF THE CAVE OF CASTLETON,
VULGARLY KNOWN BY THE NAME OF THE
DEVIL'S BOTTOM.

This grotto, regarded from all time as
the chief of the seven wonders of Derby-

shire, has been celebrated by several poets. But as, since the time of Homer, Virgil, and Ovid, who united great knowledge to sublime talents, most modern poets are hardly noted for accuracy in the description of physical facts, I shall not here quote any of the verses, with which the English Muses have inspired those who wished to paint in words this grand work of nature.

I prefer to tell the reader that this superb cavern has been honoured with the visits of several distinguished men of science, among the latest of whom were Sir Joseph Banks, President of the Royal Society of London and Dr Solander, accompanied by Omai, a savage from the South Sea Islands, who was received with so much interest in England, where he remained a considerable time, until, after being loaded with presents, he was generously conveyed back to his own country.

This cavern is situated at the foot of a great escarpment formed by nature on the brow of a steep-faced mountain, upon which stands an old castle, built, it is said, in the time of Edward, the Black Prince.

The principal entrance is 120 English

feet in width, and forty in height; it
forms a circular arch and has been opened
in a rock of grey, and somewhat sparry
limestone, hard enough to take a fine
polish, and containing a multitude of marine
petrifactions, among which the entrochi
predominate together with anomias of a
large species. This stone, when rubbed
against a hard body, gives a smell somewhat
like that of burnt horn, while some parts of
it, of a deeper grey and more sparry grain,
even emit so strong a smell, that it may be
classed among the stinking stones (*lapis
suillus.*)

An inhabitant of the place who gains a
subsistence by conducting strangers into the
interior of the cave, having heard of our
arrival, came to wait upon us. He first
presented each of us with a printed leaf
containing the most ridiculous and exaggerated details of the extraordinary things
which were to be shown to us, preceded by
the following short preamble; "As com-
plaints have been made of the exorbitant
sums demanded by those who shew this
cave, it is proper the public should be informed of what ought to be given, for those

who shew it pay no rent to the king. One
person ought to pay two shillings and
sixpence (about three French livres); and a
party together five shillings. These prices,
however, are not fixed; the public being at
liberty to give more or less as they choose.
J. HALL."

This J. Hall has taken in this preamble
a very clever way of taxing his customers;
but I must do him the justice to acknowledge
that he is active, obliging, anticipates every
question, and makes the queries and gives
the answers; while the most minute details
do not escape him respecting the objects
which he thinks most worthy of remark.
In short he knows his lesson, and repeats it
with a consequential gravity and sometimes
even with a sort of enthusiasm, that be-
speaks the interested attention of those who
are under his protection in this darksome
den.

A party of Englishmen joined us, and
we entered the outer porch. It is lighted
from without, is forty-two feet high, a
hundred and twenty feet wide, and two
hundred and seventy feet long. The light
was pretty strong at the entrance of this

vast vault, but gradually diminished as we proceeded inwards, or as the fore parts were broken into projections that advanced more or less from the sides. The effect is all the greater as the scene is enlivened with two work-shops, the one a rope yard, the other for making lace and tape, entirely within the vestibule.

All is life and motion in this place, which might have been expected to be so solitary. On the one side are seen young girls turning their wheels, folding up their tapes and threads, winding and singing at the same time; and on the other, men spinning ropes and twisting cables, or forming them into coils. What is still more extraordinary, there are in this subterranean recess two houses facing each other, entirely separate from the rock, with roofs, chimneys, doors, windows, inhabited by several families.

It is hardly possible to describe the effect of this scene and the impression produced by the sight of two houses built in a cavern. We were soon surrounded with several groups of young people, among whom were some very pretty girls.

As it appeared that J. Hall here yields

the pre-eminence to the ladies of the place, only reserving exclusively to himself the privilege of exhibiting the deeper sub-terranean grottoes, all these young girls now flocked to shew us the manufactures of rope and tape, as well as the inside of the two dwelling-houses; after which they made us admire the beauty of the vestibule, the height of the vault, and the curtains of stalactite by which it is decorated.

They called our attention more par-ticularly to a stalactite of an extraordinary shape at the beginning of the principal arch, fifty paces from the entrance, a little beyond the farthest house, and thirty-five feet high. "See," said they, "the famous *leg of pork*; notice how fine it is, and admire the per-fection of this master-piece of nature." But the more we examined this pretended "leg" the more did we find it to resemble an object which young girls are scarcely allowed to examine, and still less to allow others to examine. It was a heart and not a leg; but a heart of the same kind as those of M. de Boufflers. This shews how young folks may be persuaded that they see what they do not see ; but it was at the same time a proof of their amiable candour and innocence. Decency, however, demanded that we should preserve a grave and serious countenance, and after acknowledging our indebtedness to them by a small present, we took leave of them, and proceeded under the auspices of Hall, who, after handing to each of us a lighted torch, opened the door of a subterranean gallery at the bottom of the grand vestibule, and desired us to follow him through the darksome labyrinth of which he bustlingly held the clue.

The track seemed to us at first neither pleasant nor easy. In some places we could hold ourselves easily upright, whilst in other parts the roof was so low, that we had to walk in a stooping posture, to avoid wounding ourselves against the protuberances of the rock. This first gallery is a hundred and fifty feet long. Here we noticed a quantity of sand piled into a little, low, oblong dune. Hall, attentive to the minutest circumstances, did not omit to call our attention to this sand, which he said was the work of a small stream issuing from a subterranean pool, which we should soon reach. This stream swells after heavy rains, and carrying along with it this sand, makes the cave inaccessible during these times of overflow.

Our guide, with expressive gestures as he walked along, entertained us with an account of the rapidity of the current, the height of the water, its quality, and the noise it made; when a small lake, with a skiff floating upon it, all at once interrupted our progress. This lake, only three feet deep, is wholly inclosed in the solid rock, and stretches under a very low vault of which we could not see the farther end. Here it was necessary to stop.

We stood on the brink, and the light of our funereal torches, which emitted a black smoke, reflected our pale forms from the bottom of the lake; we almost seemed to see a troop of shades emerging from a deep abyss to come before us. The illusion was extremely striking.

This piece of water is here forty-eight feet broad; J. Hall gave it the name of the first water. He informed us, that we must cross it one by one in the little skiff, stretching ourselves at full length, as we had to pass under the arch which is very low and narrow; he assured us, however, that there was absolutely no danger.

Count Andreani would embark first; he lay down flat in the bottom of the little boat which was furnished with some straw. The guide then entered the lake and bending his head almost to the surface of the water, pushed forward the skiff with one hand, while he carried his torch in the other.

Five minutes sufficed for this crossing and the return for another passenger. My turn having come, I lay down on my back with my head depressed inside the boat; but as I passed through the low, narrow tunnel I wished to turn round on my side in order to examine the stone of which it is composed, when my hat was whisked off by the roof, and tumbled into the water. I was safely landed on the other bank, where we silently waited the arrival of some new companions.

It is impossible, however cheerful one's nature may be, not to see in this scene a picture of the passage of the dead in the fatal bark. The whole party being now assembled, Hall, after first drying himself a little, and warming his inside with a glass of rum, which he drank to the health of the travellers, called our attention to the spacious extent of the place which we had now

entered.   We found ourselves in a cavern a
hundred and twenty feet high, two hundred
and seventy feet long, and two hundred and
ten broad.   It excites real astonishment to
find such extensive natural excavations in
the centre of the hardest rock, and one is
lost in conjecturing what has become of the
materials which must once have occupied
those vast empty spaces.

In a passage at the inner extremity of
this vast cavern, we again met with water,
which our guide called the second water.
But it is easily crossed on a platform running
along the side of this small pond, which is
only thirty feet long.

On issuing from this passage we again
found ourselves in a vast cavern.   But first
we came upon a mass of rocks, from the
summit of which the water trickles, drop
by drop, and deposits a calcareous sediment.
This projection has been transformed by the
imagination into a house, though it has
not the smallest resemblance to one, and as
this so-called house incessantly receives the
drops of water from above, the genius of
rain is supposed to have made it his habitation, and this genius has not failed to be

dubbed a giant; the place is known as
Roger Rain's.

A little beyond this we came to the grand
cavern which bears the name of the *Chancel.*
The roof here is lofty; and on the sides
various cavities are seen that resemble gothic
portals and windows. Large stalactites descend from the roof upon the prominent
parts of the rock, like drapery or curtains,
and produce a very striking effect. The
pavement also is pretty smooth, being
formed of solid rock, covered over from time
to time with stalagmite. One seems to be
in an immense gothic church.

As we advanced, our conductor made a
sign to all with his hand, and a gesture to
keep silent, as if to inspire us with respect;
and he particularly desired each of us, in a
very low voice, not to look behind until it
was time and he should tell us. He then
assembled his company in a group, and
placing himself at our head with his face
towards us, continued to walk backwards, as
if teaching us the military exercise. He
kept making signs and gestures in order to
attract our whole attention, and continually
requested us to keep our eyes fixed on

himself, lest any one should be tempted to look behind. Having in this manner reached almost the inner end of the cavern, he halted his flock. We then heard sweet harmonious voices coming apparently from the lofty roof. Involuntarily turning round to see whence these angelic sounds proceeded, we observed in a natural niche at the other end, about forty-eight feet above the floor, five figures in white, motionless as statues, holding a torch in each hand, and singing in parts a fine, melodious air in some words of Shakespeare's.

We saw that Mr Hall was playing off his grand machinery for our entertainment; he was delighted at our surprise and exulted in our astonishment. This unexpected scene, indeed, made a very vivid and, at the same time, a very pleasant impression on us; it had a touching and melancholy effect that arose perhaps less from the air and the words than from the deep and remote place which separated us from the rest of nature. The ancients knew what they were about when they selected such caverns for their initiations; they never celebrated their grand mysteries save in subterranean places.

After having listened to our chantresses,
who suddenly disappeared, we resumed our
progress and marched forward through a
long gallery.   We had just been listening to
angels; we had now to take a little turn in the
infernal regions.   Our master of ceremonies,
J. Hall, introduced us into what is called *the
Devil's Cellar.*   Here we saw a great
number of names inscribed on the walls; I
know not if those who wrote them have made
a pact with the evil spirit, and if in recognition
he has made them drink all the wine in his
vaults; but this much is certain, that the
cellar is at present very ill provided.   However, as one never enters a tavern without
drinking, especially in England, Mr Hall
pulled out his little bottle from his pocket,
swallowed a glass of rum, and offered each
of us a glass after him, but we begged to be
excused, and we left this gloomy recess,
which is only a huge hole blackened by the
smoke of lamps and torches.

Hardly had we quitted the place, when we
suddenly came to a hillock of quartzose sand.
Here we had to descend by a steep path a
hundred and fifty feet long, and sinking to
the depth of more than forty feet.   By the

side of this sandy path, and for the whole of its length, there runs a deep channel, hollowed by nature in the solid rock, through which a large stream of water, having its source in the more distant parts, gently murmurs along, until it loses itself at some holes into which it plunges with loud noise.

Here J. Hall played off upon us one of the little tricks of his trade. He told us in an emphatic tone, that this subterranean brook, notwithstanding the absence of all light, produced fish, but to be sure, black fish, and to prove his assertion, he went down to the water by a narrow passage, and after plunging his hand several times into the stream, held up to our view, at some distance, one of his black fishes. He was about to throw it back into the water to prevent the destruction of its species, which according to him was beginning to become scarce, when, upon approaching to take a nearer view of it, I easily recognized it to be a tadpole, which he had carried with him to deceive us, and which was already half dead. He was himself the first to laugh at the cheat, and he frankly confessed it, as soon as he saw that it was detected.

Proceeding forward, we soon passed under
the *Arcades*, so called because the rock here
forms three different vaulted roofs, disposed
in axes of a circle like the arches of a
bridge.

A little beyond this we heard in the
distance the noise of a cascade, and saw
a pyramidal mass of stalagmite, which bears
the name of *the Tower of Lincoln*. It was
here that the cavern was formerly supposed
to end; but, a few years ago, a new gallery
was discovered, which extends four hundred
and ninety-two feet farther. This we traced
to its inmost extremity, where the little
river again appeared to our view, issuing
from a natural tunnel, as perfectly made as
if it had been a work of art, but so strait
and low, that there was no possibility of
penetrating into it. At the entrance of this
sort of aqueduct we saw several names
engraved in the rock, among which we
distinguished those of Sir Joseph Banks and
Dr Solander, and also that of Omai, who
accompanied these savants in this subterranean journey.

The entire length of the cavern, from its
entrance to the place where these inscriptions

are, is at least two thousand seven hundred and forty-two feet.

Our visit, which lasted several hours, was accomplished without the slightest accident, and we returned equally safe. We made a liberal acknowledgment for the services of our guide, who was much more tired than we were, for he had been incessantly active, discoursing and exerting himself. He was as well pleased with us as we were with his zeal and goodwill; and though he was a little chagrined at the discovery of the black fish, we took leave of each other without bearing any grudge.

Fluor Spar

Fluor-spar is an important article of production at the lead mines of Castleton, in which it is found in greater abundance than any where else. The violet is the most common kind, and serves as salband to the white sort. There are also found varieties of a fine yellow topaz colour, violet-blue, violet-purple, white inclining to rose colour, water-clear, &c. Some specimens showing a mingling of these colours have a very agreeable effect.

Fluor-spar would be the most beautiful of all substances, if it were only harder. This stone not only forms here a considerable object of traffic in its rough state, between Castleton and Derby, Winster, Matlock, Buxton, and other places in the district, but is also worked on the spot into vases and other articles of ornament, which are sent to Birmingham, where they are mounted in gilded copper or other metals.

The largest pieces of fluor-spar hardly exceed a foot in thickness, and these are by no means common.

Lead mines

The lead mines of Castleton are not very rich, and only sixty persons are employed in them; the principal productions of these mines are the different kinds of fluor-spar.

Several mines have been opened in the very steep calcareous mountain of *Mam Tor.*

Oden mine, a working not far from the town, presents one of the most extraordinary mineralogical phenomena, consisting of a specular galena, here called *slicken-sides,* which is found in large quantity, usually forming a double vein. Each vein is separated by an interval only a few lines broad, which is filled with a very white and heavy gypseous earth, to which the workmen have given the name of *keble* or *caulk*.

In order to detach large pieces of this brilliant galena, the miners make use of a sharp iron wedge, which they forcibly drive with a hammer into the thin layer of keble that separates the two veins.

When this operation is completed, the workmen retire with haste to a distance; and a few minutes after the whole vein breaks off, with a terrible noise and a general concussion, which would overturn all the props of the mine, if they had not carefully provided against such accidents by strengthening the principal beams with mine-rubbish, and leaving no empty spaces. The miners assert that a dull noise, the precursor of the explosion, marks the moment when they must beat a speedy retreat to get into shelter from accidents.

This terrible phenomenon takes place also in the mine of *Lady Wash* near the village of *Eyam*, in the same district of Derbyshire.

Mr Whitehurst has given a good description of all the circumstances connected with this phenomenon. Mr Ferber, who has likewise mentioned it, says, that no reason could be assigned to him for so extraordinary an effect: but he thinks that it may arise from the air seeking to escape, because it is strongly compressed, especially in the narrowest parts of the vein.

But to be in a position to decide in a matter of so much difficulty, it would be necessary to watch and follow with attention all the circumstances that precede or accompany the explosions, to know whether there be any combustion, and if so, whether or not it gives off any smell. It would be proper also to analyse with the most scrupulous accuracy the substance of the gangue, and the keble, which is not yet sufficiently known.

The theory of gases might above all throw great light upon this phenomenon. It is known that phosphoric acid is found in union with lead. The effects of inflammable gas mixed with phosphorus are likewise familiar, and it is known that by mere contact with atmospheric air it kindles with such rapidity that the most violent explosions may result, especially if vital air is present in large proportion. This branch of science is now sufficiently advanced to enable an intelligent observer, who should have an opportunity of tracing on the spot all the circumstances connected with this astonishing phenomenon, to give a satisfactory explanation of it.*

Toadstone containing Lead-ore

Mr Whitehurst and Mr Ferber have declared, that in the mines which have been worked, the vein is exclusively in the limestone, and disappears so completely on reaching the bed of toadstone, that not the smallest vestige of ore is discoverable in the latter; but that after piercing through the toadstone, however thick it may be, the vein never fails to reappear, and so on from bed to bed. This arrangement, most astonishing as it is, must be admitted to be in general

* [This subject has been discussed in more recent times, together with other similar phenomena in various parts of the world. The explanation that now seems most probable regards the rocks as in a state of molecular strain, like Rupert's Drop or toughened glass, and that this condition of strain is the result of the earth-movements that produced the slickensides. See A. Strahan, *Geol. Mag.*, 1887, p. 400, also pp. 511, 522.]

true and constant; it was this fact that
suggested to Dr Whitehurst the idea that the
toadstone which thus separates the calcareous
strata and cuts the veins must be the result
of different currents of lava. My way of
looking upon this subject has been already
explained, but if there should still remain
any doubt that the toadstone is not a product
of the fire of volcanoes, the fact which I am
now going to state will be sufficient to
remove it.

Dr Pearson having spoken to me at
Castleton of a miner named Elias Pedley,
who sells choice specimens for the cabinet,
we went to pay him a visit. I purchased
from him a collection of the most interesting
minerals of Derbyshire, and some fine
specimens of fluor-spar, the crystals of which
were in the most perfect preservation.

In the course of conversation with him,
I asked whether it was true, that no vein of
ore had ever been found in the toadstone.
He told me that such had uniformly been
the fact hitherto, and though himself long
engaged in the work of the mines, he had
never heard that the slightest appearance of
lead had been found in that stone; but that

he had just learned to his cost that the rule was not without exception, if not in respect to the toadstone, at least as to the cat-dirt or channel.

On my requesting a further explanation from him, he replied, that he had been ruined by working, on his own account, a vein which at first had the most promising appearance, but that, after opening a deep gallery, at a great expense, he had lost the vein, and though he found it again in the *Channel*, it was so poor that it would not repay the cost of working.

As the mine was but a little way off, he offered to shew it to us, especially when he perceived that I doubted his account. Providing himself therefore with some mining implements, he desired us to follow him, and we willingly complied.

We directed our steps about a mile to the east of Castleton, following the steep side of the hill which fronts it, and upon a narrow road about 200 feet above the plain. The hill is limestone, and in some parts exhibits traces of strata; but its general disposition presents a uniform and continuous mass like most calcareous rocks of great thickness.

Marine bodies are not very abundant here; I observed, however, a few fragments of entrochi and some terebratulæ. The hill has been pierced by some lead-mines, and it also affords calamine in an ochreous form.

We soon reached the entrance of the gallery which is horizontal. It has been opened in bedded limestone, and in a vein of white calcareous spar, which presents a small but very distinct vein of galena mixed with fluor-spar.

This indication, which was regarded as promising in a hill where there were other lead mines, determined Elias Pedley and some other associates to attack the vein. But scarcely had they reached a depth of twelve feet when the limestone ended, and they had the misfortune to meet with the channel.

As, till then, there had never been any instance of the slightest metalliferous vein being found in this barren stone, they would have immediately stopped their labours, had not the same vein of galena, which they traced through the limestone, continued its course in the channel, that is to say, in the trap. This seemed so extraordinary and so novel a thing, that seduced by it, the miners pursued the ore in the channel to the horizontal depth of ninety feet, always hoping that the vein, which never exceeded an inch in thickness, would get larger.

But the farther they went the trap became increasingly hard, and involved so much labour and expense, that Elias Pedley told us, he was on the point of abandoning the work. The trap in which they had followed the vein of lead is no more than seven feet thick, but as they had pierced it horizontally and had cut the gallery along the length of the bed, it is probable that this bank of trap runs a great way into the hill, seeing that they had already gone ninety feet in an horizontal line without discovering any indication of the least change.

This bed of channel, or cat-dirt, is really a greenish trap, very hard in the mine, but when the blocks taken out of the gallery have been exposed for some time to the air, the stone becomes friable, its colour changes, and it passes

into an earthy condition. It is probable
that this decomposition arises from some
invisible particles of pyrites, which efflo-
resce and cause the substance to fall into
mere detritus.

It is thus evident that galena exists in
this bed of cat-dirt, and it must run on for
some distance, since it has been traced without interruption for ninety feet; this vein
is accompanied with a little calcareous spar
and fluor-spar. Here then is an express
exception to what has heretofore been
observed in the mines of Derbyshire. The
trap is therefore not the work of fire, since
there is found in it a vein of lead. I am
aware that mineralogists conversant with
the study of lithology, who have examined
the trap upon the spot, and are well acquainted with this stone as well as with all
its varieties, have no need of this proof.
But the fact appeared to me to be of sufficient importance to deserve mention here,
in order to do away with every doubt upon
the subject. This consideration, therefore,
will form my excuse to those who may be
displeased with the minute and tedious
details into which I have been obliged to

enter, that I might place this question in the
clearest light.*

* [Faujas' conclusion that because a vein of galena had
once been found traversing a bed of toadstone, therefore the
toadstones could not have had a volcanic origin, is a curious
non sequitur. He did not realise that if, as is obvious, the
metalliferous veins arose long after the deposition of the rocks
through which they run, their presence affords no conclusive
proof or disproof of the origin of these rocks. There can be no
doubt that the conditions for the infilling of metalliferous veins
were much more favourable in the limestones than in the toadstones; and this difference may have been determined by the
chemical constitution of the two kinds of rock.]

CHAPTER XIX

*Derby.—Richard Brown, dealer in objects
of Natural History.—A Manufacture of
Vases, and other works in Fluor-Spar*

SATISFIED with all I had seen at
Castleton, I left that little town, and
returned to Buxton, where I put in order the
collection I had made of the most remarkable
curiosities of Derbyshire. We then waited
upon Dr Pearson, to thank him for the
kind attention he had paid us, and left Buxton next day for Derby. We were eight
hours on this journey, though we had excellent horses and good postillions, but the
roads are very bad.

Derby is a commercial town. We saw
several interesting workshops; such as por-
celain works and potteries. We had been
told of a person named Richard Brown who
resided here, dealt in objects of natural
history, and had the finest productions of
Derbyshire, as well as minerals from different

parts of England and Scotland.   We visited
his shop which is well stocked with vases
of every form and every size, as well as
other works in fluor-spar of different colours,
but much better worked and of a finer polish
than those sold at Buxton and Castleton.
I purchased from him a complete collection
of all these spars cut into small square
tablets, so that they can be placed in drawers,
which is the most convenient way of keeping them for study, and for the arrangement
of a cabinet.

We had been told that Mr Brown was
very dear, but I found that his goods, which
are more artistic and better finished than
anywhere else in the region, were at a very
reasonable price; and that so far from taking
advantage of us because we were foreigners,
he was, on the contrary, moderate in his
demands, and treated us with the greatest
civility.   When he saw that I was fond of
stones, and that I named some as to which
he had been in doubt, he loaded me with
attention; would by all means detain us to
drink to the lovers of Nature, and instantly
ordered glasses and several kinds of wine to
be brought; but as we had just left the

table, we could not accept this invitation, of the kindness of which however we were very sensible, for Mr Brown offered it with the greatest cordiality.

While we were disputing this point of politeness with him, I was cleverly robbed in the street of a dog which I had bought in the Highlands of Scotland. It differed from the common shepherd's dog. This Scottish species has at least as much intelligence and capacity as the other in the management of a flock of sheep; it is besides, very good for hunting the fox. I tried in vain to recover the animal: my dog was lost or rather stolen.

Next day we went to see another natural history dealer, who was himself a worker in fluor-spar and marble. He lives at one of the ends of the town, by the side of a small stream which flows at a short distance from his house. He is a most intelligent young man. Nowhere have I seen vases so perfect in shape, so light, and made of such choice materials; but he is dearer and much less accommodating than Mr Brown. Notwithstanding this, as I wished to take something from him, I bought a vase which charmed me by the beauty of the colours of the spar, by its form, and by the finish of its workmanship. This dealer sells also some of the minerals of Derbyshire, but I found nothing very interesting here in that kind.

CHAPTER XX

Departure from Derby.—Arrival at Bir-
mingham ; its numerous Manufactures.—
Doctor Withering. — James Watt. —
Doctor Priestly ; his House ; his Library ;
his chemical Laboratory

WE left Derby at noon, but as the roads
are all still very bad along the whole
of this route, we had much difficulty in
arriving on the same day at Birmingham.
It was past nine in the evening when we
reached the inn, after having crossed black
and arid heaths, and an extremely wild
country.

We had letters of recommendation to
Doctor Withering,* the translator of the
Sciagraphia of Bergmann, and a lover of
both botany and chemistry : we hastened to
wait on him next day to present them. He

* [William Withering (1741-1799) studied medicine at
Edinburgh, where he graduated ; chief physician to the Birmingham General Hospital ; noted for his acquirements in botany
and mineralogy ; elected F.R.S. in 1748.]

lives in a fine house, furnished with much
taste and elegance.   We had tea at his house
in company with ladies as pleasant as they
were good-looking, and to complete our
good fortune, we were here introduced to
Mr Watt, one of the most scientific men in
England as regards mechanics, and who also
possesses great knowledge in chemistry and
physics.*

From the activity of its manufactures and
its commerce, Birmingham is one of the most
curious towns in England. If any one
should wish to see in one comprehensive
view, the most numerous and most varied
industries, all combined in contributing to
the arts of utility, of pleasure, and of luxury,
it is hither that he must come.   Here all
the resources of industry, supported by the
genius of invention, and by mechanical skill
of every kind are directed towards the arts,
and seem to be linked together to co-operate
for their mutual perfection.

I know that some travellers who have not
given themselves the trouble to reflect on
the importance and advantage of these kinds
of manufactures in such a country as

* [James Watt (1736-1819), the illustrious engineer.]

England, have disapproved of most of these industrial establishments. I know that even an Englishman who has only taken a hasty, I would almost say an inconsiderate view of these magnificent establishments, William Gilpin, has said that it was difficult for the eye to be long pleased in the midst of so many frivolous arts, where a hundred men may be seen, whose labours are confined to the making of a tobacco box.* But besides that this statement is exaggerated and ill-considered, its author has not deigned to cast his eyes over the vast works where steam-pumps are made, these astonishing machines, the perfecting of which does so much honour to the talents and knowledge of Mr Watt; over the manufactories in constant activity making sheet-copper for sheathing ships' bottoms; over those of plate-tin and plate-iron, which make France tributary to England, nor over that varied and extensive hard-ware manufacture which employs to so much advantage more than thirty thousand hands, and compels all Europe, and a part of the New World, to supply themselves from England, because all

* See Gilpin's " Picturesque Tour in England."

ironmongery is made here in greater perfection, with more economy and in greater
abundance, than anywhere else.

Once more, I say with pleasure, and it
cannot be too often said to Frenchmen, that
it is the abundance of coal which has performed this miracle and has created, in the
midst of a barren desert, a town with forty
thousand inhabitants, who live in comfort,
and enjoy all the conveniences of life.

Here a soil, once covered with the most
barren and sombre heath, has been changed
into groves of roses and lilacs, and turned into
fertile and delightful gardens by Mr Boulton,
associated with Mr Watt, in whose work
more than a thousand hands are engaged.

The population of Birmingham has made
such an advance, that during the war with
the United States of America, a war which
weakened the resources of England, at
least three hundred new houses were added
annually to the town, and this rate doubled
as soon as peace was concluded. A well-
informed person assured me that this was
true, and he showed me, during my stay
in the town, a whole street which was in
process of erection with such rapidity that,

all the houses being built on a given plan and at the same time, one could believe that the street would be entirely completed in less than two months.

I had much pleasure in visiting Mr Watt as often as I could, whose wide accomplishments in chemistry and the arts, inspired me with the very greatest interest. His moral qualities, and his engaging manner of expressing his thoughts, increased my respect and regard for him. He joins to the frank manners of a Scotsman the gentleness and amiability of a kind - hearted man. Surrounded with charming children, all well-informed, full of talent and having had the best education, he enjoys among them the unalloyed happiness of making them his friends and of being adored by them as their father.

I had one day a delightful dinner with this amiable family, and it was doubly interesting to me, for Doctor Priestley * who

* [Joseph Priestley (1733-1804), one of the most illustrious men of science of his time ; first a presbyterian clergyman, and author of writings on philosophy and theology; he was more particularly distinguished for the originality and insight of his contributions to chemistry and physics. Finding his social and religious views highly unpopular in England, he emigrated to the United States in 1794, and settled in Pennsylvania where he died.]

is a relation and friend of Mr Watt, was
there; I had the pleasure of making the
acquaintance of this celebrated man, to
whom the physical sciences owe such great
obligations, and whose gentle and kindly
manners increase our affection for his virtues.

In company with Dr Withering, I paid a
visit to Dr Priestley, who does not reside in
Birmingham itself, but about a mile and a
half from the town, in a charming house,
with a fine meadow on the one side, and a
delightful garden on the other. The most
perfect neatness pervades everything, both
without and within this house. I know not
how to give a better idea of its site and
construction than by comparing it to those
pretty and elegant Dutch houses, distributed
in such profusion on the road from Harlem
to Leyden, or from Leyden to the Hague.

Dr Priestley received me with much kindness. He presented me to his wife and his
daughter, who both have as much vivacity
of mind as gentleness of manner. The
young lady spoke to me of one of her
brothers, who was then finishing his education at Geneva, and to whom she seemed to
me to be strongly attached.

Dr Priestley's chemical and physical laboratory stands at the end of a court, and is detached from his house to avoid the risk of fire. It consists of several rooms on a ground floor. Upon entering it, we were struck with a simple and ingenious apparatus for making experiments on inflammable gas, extracted from iron and water reduced to vapour. The tube, which was thick and long, was made of copper and cast in one piece to avoid joints. The part exposed to the fire was thicker than the rest. Into this tube he introduced filings or slips of iron, and instead of dropping in the water, he preferred making it enter as vapour. The furnace destined for this operation was heated with coke made from coal, the best of all fuels for the intensity and equality of its heat.

By these means he obtained a considerable quantity of inflammable gas of great lightness and without any smell. He observed to me, that by increasing the apparatus and using iron or copper tubes of a larger calibre, aerostatic balloons might be filled at small expense and without the trouble and cost involved in the use of vitriolic acid. Dr

Priestley allowed me to take a drawing of this new apparatus for the purpose of communicating it to the French chemists who are working at the same subject.

The luting which Dr Priestley used to prevent the gas from escaping, either in this or other experiments connected with gases, appeared to me so good that I begged of him to tell me its composition. He told me, that after a multitude of trials, he had found nothing better than the paste of almonds, such as it is when the oil is extracted. This moistened with a little water, in which glue has been dissolved, makes an excellent impermeable lute. He added, that the glue might be dispensed with.

Dr Priestley did not yet regard the experiments on the decomposition of water as wholly satisfactory. He could not admit the fact to be demonstrated so long as the gas was only obtained through the medium of iron, a metal which is itself capable of being burnt; but he waited with impatience for the result of the experiments of the French chemists, particularly of Lavoisier, who had invented, and caused to be constructed, a splendid apparatus for this object.

"The decomposition of water," said this indefatigable philosopher to me, "is of such high importance in physics, and would play so great a part in most of the phenomena of nature, that far from admitting the fact upon slight evidence, and as it were from enthusiasm, it is rather to be wished that all possible objections that may be made, and will still long continue to be made, should be completely refuted. It is from the conflict of opinions, that truth at last emerges victorious. But I have still so many doubts, and I have so many experiments to make, or to repeat *pro* and *con*, that in spite of all that has been done up till now, I regard the question as only blocked out." *

Dr Priestley has adorned his solitude with

* Mr James Watt, who has published some excellent papers upon the theory of fire, was of the same opinion with Dr Priestley. "The theory of the decomposition of water is seductive," he said to me, "for it would be convenient for explaining different phenomena of nature ; but the more I reflect on this delicate subject, and upon all that has hitherto been done and written relative to it, the more do I see difficulties arise." Some years have passed since that time and the war has interrupted our relations with England ; but we heard two years ago, by way of Hamburg, that Messrs Priestley and Watt were still of the same opinion. It is possible, however, that since then they have come to agree with the French chemists.

a cabinet of physics which contains the instruments necessary for his experiments, and a library rendered valuable by a choice of excellent works. The learned possessor employs himself in a variety of studies. History, moral philosophy, and religion, have all in turn engaged his pen. His active mind, his powers of observation, and his natural thirst for knowledge, lead him to love the physical sciences; but his gentle and kind-hearted disposition has sometimes drawn him into pious and philanthropic ideas, which do honour to the feelings of his heart, since the motives that actuate him always have for their object the happiness of mankind. Besides, his position as doctor of divinity makes it often necessary for him to speak in public.*

* I do not wish to recall here the details of the causes of the persecution which this worthy man experienced since the period at which I saw him. His chemical laboratory, his cabinet of physics, his library, his charming house, were all destroyed by barbarous hands. The government has endeavoured to repair this loss by proportional indemnities, which amounted to fifteen thousand louis. But as a philosopher, who should fly from intrigue and seek repose, avoiding the recurrence of such risks, he has retired to the United States of America. Free in that asylum to follow his tastes and continue his labours, let us hope he may be able to return to them with the same zeal, and that he may repair, at least in part, the loss of his latest manuscripts.

It was a great pleasure to me to see this estimable philosopher in the midst of his books, his furnaces, and his physical instruments, surrounded with his family, a well-informed wife and an amiable daughter, and in a delightful home where everything breathed peace and goodness.

Next day I had again the pleasure of meeting Dr Priestley at Mr Watt's, where we had a pleasant repast, in company with several amiable and intelligent men. Mr Watt is a man of large conceptions. Nature has endowed him with an excellent judgment and a strong head, and to such happy qualities he joins a gentle and winning character which makes him most engaging and well deserving of the affection of all who meet him even for the first time.

Mr Watt shewed us a corn-mill, which he had just had constructed, and which was set in motion by a steam engine. The application of this principle to the mechanism of a mill is a happy idea, which may be very advantageously applied in a country with a scarcity of water, but rich in coal. This first attempt will lead to others, and a multitude of factories will soon be set going on the same principle.*

Mr Watt is so familiar with great inventions, so versed in the higher branches of mechanics, and has so highly perfected the means of execution, that he may justly be ranked among the men who have most contributed to the prosperity of the useful arts and of commerce in England. He is a Scotsman by birth. Scotland has long been able to supply England with men who honour it the most in every way.

We passed several days at Birmingham in

* Since this period similar mills have been successfully erected at Nantes, and some at Paris, where some steam engines are even used for stamping coins. Steam engines were first established in France, by the brothers Periers, who joined much activity to a great deal of knowledge. But these excellent machines cannot reach the perfection they have obtained in England, until our government shall seriously turn its attention to the opening of coal-mines, and not wait until compelled by necessity. Those who know the exhausted state of our forests are convinced that this moment will soon arrive. Those who are acquainted with the neglected situation of our coal-mines, tremble, lest it should hereafter happen that we shall want both wood and coal at the same time. There is some reason for anxiety on this subject, when we consider that at Paris a cart-load of bad coal costs six times more than the best did some years ago. But if this subject were taken into serious consideration as well as that of canals, it would be found that our resources of this kind will be as inexhaustible as all those of which we have given proof to the great astonishment of all Europe.

the midst of the arts and industries, and in the society of enlightened men and amiable women. Nothing can equal so peaceful a charm; the mind is fed and inspirited; the head is filled with facts, and the heart with gratitude. Such was our experience in this town which we could not leave without regret.

CHAPTER XXI

Departure from Birmingham.—Coventry.—
Warwick. — Oxford. — Saint Albans.—
Barnet.—London.—Return to France

AS we were preparing to leave Birmingham, Mr Watt requested to know
whether we could take under our care one of
his sons, who was to go to Paris, and thence
to Geneva. We answered, that we should
do so with much pleasure, that we would
give him a seat in our carriage, which was
sufficiently commodious, and that we should
take every possible care of the young man
who himself was very engaging. Count
Andreani and myself were delighted to
respond to Mr Watt's confidence, and to
show him how happy we were to have it
in our power to give him this trifling mark
of our esteem and attachment.

Next day we took the road for London.
On leaving Birmingham it was pleasant to
see the country on every side sprinkled over

and adorned with charming country houses,
simple but elegant, and set off by the effect
of the rosy colour of the bricks upon the
white ground of the stone-work. Every
thing here was so much more fresh, seeing
that these pretty habitations are almost new.
But scarcely had we lost sight of them, and
passed through some wooded parts when we
entered upon bare country full of heaths, the
aspect of which is as wild as it is barren.

Between Birmingham and Coventry we
had a view of an ancient mansion belonging
to Lord Aylesford. Its situation is not
pleasant, but one could see that the proprietor
had called in the assistance of taste and art
in embellishing it.

Coventry is a pretty little town. The
spire of the church is seen a great way off.
The soil here consists of rolled pebbles
mixed with reddish earth.

From Coventry to Warwick we passed
over a flat country, sometimes wooded,
sometimes bare, with a soil of the same
kind.

We stopped at Warwick in order to visit
the church, which is a very fine structure in
the gothic style. The chapel where the

chiefs of the house of Warwick are interred, and which contains the tomb of the earl of Leicester, is charged with sculptures and gothic ornaments of great finish. After visiting the other curious monuments of the place, of which long descriptions are to be found in various works, we went on to Stratford, celebrated as the birthplace of the immortal Shakespeare. We crossed the river Avon by a bridge of fourteen arches, erected at the expense of Hugh Clifton, who was a mayor of London, and a native of this town.

We next reached Oxford, where we visited the most remarkable monuments of science and art. But all these are already so generally known, that it would be useless to speak of them here. I should have been happy to meet here Mr Thompson, a very good naturalist, whose acquaintance I had the advantage of making in London, and who had gone to settle at Oxford. But he was unfortunately absent from home. It would have been not only very agreeable, but also very useful to us to have seen him, as he could have introduced us to the learned men of the place, to whom, relying

upon finding him here, we had not provided
ourselves with any letters of recommendation.

From Oxford we went on to London
through Saint Albans and Barnet.

Our stay in London was not long, for
after having taken leave of the learned
friends, who had kindly gratified us with
objects of instruction and entertainment,
which that city affords in such abundance,
we set out for Paris, where we arrived five
days after.   Count Andreani prepared for
his return to Milan, young Watt took the
road to Geneva, and I remained at Paris.

POSTSCRIPT

*T*HIS *Journey was in the press in the second year of the Revolution. Its progress had to be stopped in the midst of that most terrible of commotions. But now that the law has resumed its sway, the sciences will soon follow in its train. If these have not in England suffered as long and continuous a stagnation as in France, they have at least been reduced there to a languishing condition, in which their advance has been by no means rapid. The English, however, have had the good fortune to lose none of the men who have shed lustre on their country in the domain of polite learning, while we have had to grieve over the assassination of a large part of ours. I hope, therefore, that the reader will pardon the sad and perhaps querulous tone that has crept into some of the notes which I have appended to this work. I wish cordially and sincerely to forget all the evil that has been done to myself, but it is impossible for me to keep silence on what has been done to others.*

INDEX

Heather in Highlands, I. 233,
II. 91, 155.

Hebrides; use of term, II. 2;
stormy-seas of, 17, 20, 22,
28; boatmen of, 27; boat-
songs of, 29, 35.

Hematite, II. 149.

Hemp, Chinese, I. 13.

Hérault de Séchelles, I. 114

Herschel, Caroline, I. 63.

Herschel, William, I. 54, 56,
60, 63-77.

Highland garb, I. 202, 264,
289, II. 13, 14, 66;
cottages, I. 229, 232, 290,
II. 75; hospitality, I. 240-
255, 263, 292, 320, II. 10,
16, 21, 30, 87, 89, 92, 93;
innkeepers, I. 229, 232, 238,
263, 267, 276, II. 88, 96,
156, 159, 172; school-
master, I. 267, II. 150;
manuscripts, I. 286, 287;
weapons, I. 288, 289, II.
14; pine-lamp, I. 293;
sensitiveness, I. 297; roads,
I. 237, 261, 299, 300, 303,
307, II. 13, 88, 94, 154;
fuel, I. 292, II. 38, 153;
bagpipes, I. 319, 321, II.
247, 250; politeness, I.
321; boats, II. 3; horses,
II. 11, 78, 88; vermin, II.
19, 22; navigation, II. 3, 6,
17, 27; boat songs, II. 28,
29, 35; love of home, II.
64; tartan, II. 67; kinds of
bread, II. 67; breakfast, II.
68, 69; dinner, I. 249, II.
70; winter, II. 74; crops,
II. 74, 76, 170; bare heads
and feet, II. 75; diet, II.
76, 77; climate, II. 76,
77, 81; longevity, II. 77;
cattle, II. 78, 79; sheep, II.

78, 80-86; hostelries, I. 229,
231, 238, 263, 303, 312, II.
88, 96, 99, 103, 156, 170,
172, 180; woods, cutting
down of, II. 150; peat, I.
292, II. 152, 155; music, I.
319, 321, II. 247, 250;
equality of rank in musical
competitions, II. 251.

Highwaymen in England, I.
61, II. 271.

Hill, Professor George, II.
193, 195.

Hill, T. F., "Ancient Erse
Poems" of, I. 290.

Home, Professor Francis, II.
226.

Hope, Professor John, II. 226,
228.

Hornstone, I. 222.

Hotel-bills, instances of ex-
tortionate, I. 195, II, 270.

Houston, William, I. 4.

Howard, Dr, II. 232, 234.

Hume, David, II. 256.

Hunter, John, I. 38.

Hunter, Professor Andrew, II.
225.

Hunter, Professor John (St
Andrews), II. 195.

Hutton, James, I. 191, II. 228,
234.

ICELAND, I. 82.

India-rubber tubes, early form
of, I. 31.

Inverary, I. 237, 242, 261;
Castle, II. 240, 243.

Inverkeithing, I. 195.

Iona, isle of, II. 17, 21, 63.

Iron-ore, I. 185, II. 149.

Iron-works, I. 178, II. 148.

Islay, dyke in, II. 144.

Italians in Mull, II. 100.

JARDIN des Plantes, Paris, I. 32.

PRINTED BY
TURNBULL AND SPEARS,
EDINBURGH

Printed in the United States
By Bookmasters